Paradoxes in Probability Theory
and Mathematical Statistics

Mathematics and Its Applications *(East European Series)*

Gábor J. Székely

Paradoxes in Probability Theory and Mathematical Statistics

D. Reidel Publishing Company

A MEMBER OF THE KLUWER ACADEMIC PUBLISHERS GROUP

Dordrecht / Boston / Lancaster / Tokyo

Library of Congress Cataloging in Publication Data

Székely, Gábor J., 1947–
Paradoxes in probability theory and mathematical statistics.

(Mathematics and its applications. East European series)
Translation of: Paradoxonok a véletlen matematikában.
Bibliography: p.
Includes indexes.
1. Probabilities. 2. Mathematical statistics. 3. Paradox. I. Title. II. Series: Mathematics and its applications (D. Reidel Publishing Company).
East European series.
QA273.S9413 1986 519.2 86–7265
ISBN 90-277-1899-7

Distributors for the United States and Canada
Kluwer Academic Publishers,
101 Philip Drive, Assinippi Park, Norwell, MA 02061, U.S.A.

Distributors for Albania, Bulgaria, China, Cuba, Czechoslovakia, German Democratic Republic, Hungary, People's Republic of Korea, People's Republic of Mongolia, Poland, Roumania, Soviet Union, Socialist Republic of Vietnam, and Yugoslavia
Kultura Hungarian Foreign Trading Company
P.O.B. 149, H-1389 Budapest 62, Hungary

Distributors for all remaining countries
Kluwer Academic Publishers Group,
P.O. Box 322, 3300 AH Dordrecht, Holland

Joint edition published by D. Reidel Publishing Company, Dordrecht, Holland
and
Akadémiai Kiadó, Budapest, Hungary

This book is the revised edition of *Paradoxonok a véletlen matematikában,* Műszaki Könyvkiadó, Budapest
Translated by Márta Alpár and Éva Unger

Printed in Hungary.

Contents

Series Editor's Preface

Approach your problems from the right end and begin with the answers. Then one day, perhaps you will find the final question.

'The Hermit Clad in Crane Feathers' in R. van Gulik's *The Chinese Maze Murders*.

It isn't that they can't see the solution. It is that they can't see the problem.

G. K. Chesterton. *The Scandal of Father Brown* 'The point of a Pin'.

Growing specialization and diversification have brought a host of monographs and textbooks on increasingly specialized topics. However, the "tree" of knowledge of mathematics and related fields does not grow only by putting forth new branches. It also happens, quite often in fact, that branches which were thought to be completely disparate are suddenly seen to be related.

Further, the kind and level of sophistication of mathematics applied in various sciences has changed drastically in recent years: measure theory is used (nontrivially) in regional and theoretical economics; algebraic geometry interacts with physics; the Minkowski lemma, coding theory and the structure of water meet one another in packing and covering theory; quantum fields, crystal defects and mathematical programming profit from homotopy theory; Lie algebras are relevant to filtering; and prediction and electrical engineering can use Stein spaces. And in addition to this there are such new emerging subdisciplines as "experimental mathematics", "CFD", "completely integrable systems", "chaos, synergetics and large-scale order", which are almost impossible to fit into the existing classification schemes. They draw upon widely different sections of mathematics. This programme, Mathematics and Its Applications, is devoted to new emerging (sub) disciplines and to such (new) interrelations as exempla gratia:

- a central concept which plays an important role in several different mathematical and/or scientific specialized areas;

- new applications of the results and ideas from one area of scientific endeavour into another;
- influences which the results, problems and concepts of one field of enquiry have and have had on the development of another.

The Mathematics and Its Applications programme tries to make available a careful selection of books which fit the philosophy outlined above. With such books, which are stimulating rather than definitive, intriguing rather than encyclopaedic, we hope to contribute something towards better communication among the practitioners in diversified fields.

Because of the wealth of scholarly research being undertaken in the Soviet Union, Eastern Europe, and Japan, it was decided to devote special attention to the work emanating from these particular regions. Thus it was decided to start three regional series under the umbrella of the main MIA programme.

Progress in mathematics, as in other sciences, thrives on the kind of questions and/or results which, so to speak, require one to twist one's mind upside down and out of shape a couple times in order to see that the solutions are both natural and beautiful. Paradoxes, i.e. counterintuitive but true results, are perhaps the purest manifestations of such problems. And probability theory, the science of random events, has always been and still is particularly rich in paradoxes.

Studying and understanding a field through paradoxes is probably one of the better ways of gaining real intuition. For probability theory this would be an ideal book to do so.

The unreasonable effectiveness of mathematics in science ...

Eugene Wigner

Well, if you know of a better 'ole, go to it.

Bruce Bairnsfather

What is now proved was once only imagined.

William Blake

As long as algebra and geometry proceeded along separate paths, their advance was slow and their applications limited.

But when these sciences joined company they drew from each other fresh vitality and thenceforward marched on at a rapid pace towards perfection.

Joseph Louis Lagrange.

Bussum, March 1986 Michiel Hazewinkel

Introduction

"The fairest thing we can experience is the mysterious. It is the fundamental emotion which stands at the cradle of true art and science. He who knows it not and can no longer wonder, no longer feel amazement, is as good as dead, a snuffed-out candle."

Albert Einstein; *Mein Weltbild*, 1934 Engl. translation: Ideas and Opinions, by S. Bargmann

"It is remarkable that a science which began with the considerations of games of chance should have become the most important object of human knowledge..."

Pierre Simons, Marquis de Laplace; *Théorie Analytique des Probabilités*, 1812

Just like any other branch of science, mathematics also describes the contrasts of the world we live in. It is natural therefore that the history of mathematics has revealed many interesting paradoxes some of which have served as starting-points for great changes. The mathematics of randomness is especially rich in paradoxes. According to Charles Sanders Peirce no branch of mathematics is as easy to slip up in as probability theory. This book aims to show how this rapidly progressing and widely used branch of knowledge has developed from paradoxes. It tries to show those exciting moments that preceded or followed the solution of some outstanding paradoxical problems which are rarely mentioned in monographs. The book deals not only with interesting but not very important "gems" of probability theory, far from the main stream of development; on the contrary it emphasizes the contradictions that have done the most to clear up fundamental crises in the mathematics of randomness. The book also deals with problems that were not originally regarded as paradoxes. A book on paradoxes must naturally have a historical framework and so this book begins with the oldest paradoxes of probability theory.

It is important to distinguish paradoxes from fallacies. The first one is a true though surprising theorem while the second one is a false result obtained by reasoning that seems correct. Both paradoxes and fallacies are very interesting and instructive but this book deals mainly with paradoxes (exceptions are, e.g., the "paradoxes" of IV/1). In formulating the paradoxes my aim was that each paradox should be clear by itself. It is obvious though that the reader who does not play bridge, or is not familiar with normal distributions, will have more difficulty when these particular notions are the basis of the paradoxes. However, by reading the book straight through from the beginning, he will discover the definitions of the most important notions. (The rules of bridge are not discussed in the book but those which are necessary to understand the paradox can also be found out.)

The book consists of four main chapters. Each paradox will be discussed in five parts: the history, formulation, explanation of the paradox, remarks, and, finally, references. Each chapter finishes with quickies. These are not discussed in detail, not because they are of less importance or interest, but because they do not fit into the main line of the book.

The initial inspiration for a book on the paradoxes of probability theory came from my late professor, *Alfréd Rényi*. *A. N. Kolmogorov* also encouraged me when we met in Budapest in 1972. In 1976, I spent a semester at the University of Amsterdam, where Professor *A. A. Balkema* drew my attention to several interesting paradoxes. Further inspirations came from the discussions following my lectures at Johns Hopkins', Columbia, Yale University and at MIT. I was also fortunate to have the opportunity to meet and discuss probabilistic problems with *George Polya* at Stanford University and in Budapest. I would like to thank him for his advice. Special thanks must be given to several colleagues of mine at the L. Eötvös University and the Mathematical Institute of the Hungarian Academy of Sciences. Their names will occur frequently.

Finally, it should be emphasized that this English edition of the book is a revised and updated version of the Hungarian one.

Chapter 1

Classical paradoxes of probability theory

"A classic is something that everybody wants to have read and nobody wants to read."

Mark Twain

"Experience is the name everybody gives to his mistakes."

Oscar Wilde

"...the true logic for this world is the calculus of Probabilities, which takes account of the magnitude of the probability which is, or ought to be, in a reasonable man's mind."

J. Clerk Maxwell

Considerations on probabilities (such as the old golden rules of gamblers) can be traced back to ancient times but mathematical calculations on probabilities and probabilitistic paradoxes have been put in writing only since the beginning of modern times. Though probability theory today has about as much to do with games of chance as geometry has to do with land surveying, the first paradoxes nevertheless arose from popular games of chance.

1. THE PARADOX OF DICE. "GAMES OF CHANCE" IN THE WORLD OF PARTICLES

a) The history of the paradox

Dice was the most popular game of chance up until the end of the Middle Ages. The word hazard refers to dice as well, for it comes from the Arabic "al-zar" meaning "the dice". Card games became popular in

Europe only in the 14th century, while dice had already been in fashion in ancient Egypt during the 1st dynasty and later in Greece, as well as in the Roman Empire. (According to Greek tradition it was *Palamedeo* who invented dice in order to entertain the bored Greek soldiers waiting for the battle of Troy. A 2nd century writer, *Pausanias,* mentions a picture painted by *Polygnotos* in the 5th century BC which showed Palamedeo and Thersites playing dice.) The earliest book on probability theory is a book by *Gerolamo Cardano* (1501—1576) called "De Ludo Aleae" which is devoted mostly to dice. This short book was published only in 1663 about 100 years after it had been written. It might have been the reason why *Galileo* began to deal with the same dice-problem, although it had already been solved in Cardano's work. Galileo also wrote a study on this theme sometime between 1613 and 1624. Its original title was "Sopra le Scoperte dei Dadi" but in the 1718 edition of Galileo's collected works, the title was changed to "Consideratione sopra il Giuoco dei Dadi".

b) The paradox

A fair dice, when thrown, has an equal chance of falling on any of the numbers 1, 2, 3, 4, 5 or 6. In the case of two dice the sum of the numbers thrown is between 2 and 12. Both 9 and 10 can be made up in two different ways out of the numbers 1, 2, ... 6. $9=3+6=4+5$ and $10=4+6=$ $=5+5$. In the 3 dice problem, both 9 and 10 can be made up in six ways. Why then is 9 more frequent if we throw two dice, and 10 if we throw three?

c) The explanation of the paradox

The problem is so simple to solve that it is really surprising that people at that time found it so shocking. Both *Cardano* and *Galileo* pointed out that the order of the cast must be taken into consideration. (Otherwise not all results would be equally probable.) In the case of 2 dice, 9 and 10 can be made up as follows: $9=3+6=6+3=4+5=5+4$ and $10=$ $=4+6=6+4=5+5$. This means that in the 2 dice problem we can throw 9 in four ways but 10 only in three ways. Therefore the chance of getting 9 is more likely. (Since 2 dice can make $6 \cdot 6 = 36$ different number pairs of the same probability, the chance of getting a 9 is $\frac{4}{36}$ while that of for 10 is only $\frac{3}{36}$.) In the case of 3 dice it is just the other way round. 9 can be obtained only 25 ways but 10 26 ways. So 10 is more probable than 9.

d) Remarks

(*i*) In spite of the simplicity of the dice problem, several great mathematicians failed to solve it because they forgot about the order of the cast. (This mistake is made quite frequently, even today.) Leibniz, one of the creators of the differential and integral calculus, and D'Alembert, one of the greatest authors of the famous French Encyclopedia, were both mistaken. D'Alembert was once asked the following question: What is the probability of a coin falling at least oncc heads if it is tossed twice? The scientist's answer was 2/3, because he thought that there were only three possible outcomes (heads-heads, heads-tails, tails-tails) and among these only one is unfavourable, i.e., when we toss two tails. He neglected that the three possible outcomes are not equally probable. The correct answer is 3/4, because tossing heads-heads, heads-tails, tails-heads and tails-tails have the same chance and only the last one is unfavourable. D'Alembert's opinion was even published in the Encyclopedia 1754 at the entry "Croix on pile".

(*ii*) The dice problem has some links with 19th and 20th century microphysics. Suppose that we play with particles instead of dice. Each face of the dice represents a phase cell on which the particles appear randomly and which characterizes the state of the particles. Here dice is equivalent to the *Maxwell—Boltzmann* model of particles. In this model (used mostly for gas molecules) every particle has the same chance of reaching any cell, so in a list of equally probable events, the order must be taken into account, just as in the dice problem. There is another model in which the particles are indistinguishable, and for this reason the order must be left out of consideration when counting the equally possible outcomes. This model is named after *Bose* and *Einstein*. Using this terminology the point of our paradox is that dice are not of the Bose—Einstein but of Maxwell—Boltzmann type. It is worth mentioning that none of these models are correct for bound electrons because in this case, only one particle may occupy any cell. In dice-language it means that after having once thrown a 6 with one of the dice, we cannot get another 6 on the other dice. This is the *Fermi—Dirac* model. Now the question is which model is correct in a certain situation. (Beside these three models, there are many others not mentioned here.) Generally we cannot choose any of the models only on the basis of pure logic. In most cases it is experience or observation that settles the question. But in the case of dice, it is obvious that the Maxwell—Boltzmann model is the correct one and at this moment that is all we need.

e) References

A classical monograph on the history of the classical probability theory is:

Todhunter, I., *History of the Mathematical Theory of Probability,* which was firstly published in 1865 and republished in 1949 by the Publishing House Chelsea.

The description of the prehistory and earliest periods of probability theory can be found in:

David, F. N., *Games, Gods and Gambling.* Griffin, London, 1962.

The following books point out the historical and philosophical aspects of early probability theory:

Hacking, I., *The Emergence of Probability.* Cambridge University Press, 1973.

Maistrov, L. E., *The Development of the Notion of Probability* (in Russian). Nauka, Moscow, 1980.

Readers interested in the history of early probability theory may find further details in the periodicals *Biometrika* and *Archive for the History of Exact Sciences.*

The English translation of the first monograph on the problems of dice as well as the detailed biography of its author Gerolamo Cardano can be found in

Ore, Ø., *Cardano, the Gambling Scholar.* Princeton University Press, Princeton, 1953.

2. DE MÉRÉ'S PARADOX

a) The history of the paradox

There is an old story (probably from Leibniz) that the well-known 17th century French gambler Chevalier de Méré was on his way to his estate in Poitou when he met *Blaise Pascal,* one of the most famous scientists of the century. De Méré posed two problems to Pascal, both connected with games of chance. The first problem was the paradox in question (the second one is the next paradox). In 1654 Pascal corresponded with *Pierre de Fermat,* another highly gifted scientist living in Toulouse, about these two questions. They both came to the same result, which pleased Pascal very much. He writes in a letter: "I see that the truth is the same in Toulouse and in Paris." *Øystein Ore,* professor at Yale University, has pointed out that the paradoxes attributed to de Méré had, in fact, been common knowledge much earlier, it was just that Pascal had not known about them. Nor is it true that the chevalier was a passionate gambler. He was interested in paradoxes theoretically rather than practically, which is why he was not satisfied that Pascal had "only" solved the problem, confirming the answer he already knew was right He could not see from the solution how the contradiction was solved.

b) The paradox

In four throws of a single dice the probability that we get at least one ace is more than 1/2, whereas in 24 throws of 2 dice the probability that we get a double ace (at least once) is less than 1/2. This seems surprising since the chance of getting one ace is six times as much as the chance of a double ace, and 24 is exactly 6 times as great as 4.

c) The explanation of the paradox

If one true dice is thrown k times, then the number of possible (and equally likely) outcomes is 6^k. In 5^k cases out of these 6^k, the dice does not turn up a six, hence the probability of throwing at least one ace in k

throws is

$$\frac{6^k-5^k}{6^k}=1-\left(\frac{5}{6}\right)^k,$$

and that is greater than 1/2 if $k=4$. On the other hand, the quantity $1-\left(\frac{35}{36}\right)^k$, which we can obtain in the same way, is still smaller than 1/2 for $k=24$ and only exceeds 1/2 for $k=25$. So the "critical value" is 4 for a single dice and 25 for a pair of dice. This undoubtedly correct solution did not in fact satisfy de Méré as he had known the answer itself; it was just that he did not understand why it was incompatible with the "proportionality rule of critical values", which says that if the probability decreases one sixth times, then the critical value increases six times (4:6=24:36). *Abraham de Moivre* (1667—1754) proved in his book "Doctrine of Chances", published in 1718, that the "proportionality rule of critical values" was not far from the truth. For if p is the probability of an event (for example, the probability of throwing an ace is $p=1/6$), then the critical value k can be calculated by solving the equation

$$(1-p)^x=\frac{1}{2}$$

(the equation can be solved if p is strictly between 0 and 1). The critical value k is the smallest integer which is greater than x. The solution of the above equation is:

$$x=-\frac{\ln 2}{\ln(1-p)}=\frac{\ln 2}{p+p^2/2+\dots}; \qquad (*)$$

where ln denotes natural logarithm (base $e=2.71\dots$). It is apparent from this solution that if p^2 is negligibly small, then p decreases approximately in proportion to the increase in the critical value, just as de Méré thought. De Moivre used the approximation formula $x\approx\frac{\ln 2}{p}\approx\frac{0.69}{p}$ to examine the Royal Oaks Lottery (the London Lottery). In that case the value of p was 1/32 and for $p=\frac{1}{32}$ the correct value is $x=22.135\dots$, while the above formula gives the approximation 22.08, which is very near to the correct value. De Méré's paradox occurred because for $p=1/6$, $p^2/2$ (and other terms in the denominator of formula (*)) are not small enough to neglect. Thus the "proportionality rule of critical values" is just an approximate rule, the error of which increases as p increases. This is the real solution of the paradox.

d) Remarks

(*i*) A typically incorrect solution of de Méré's problem goes back to Cardano. He reasoned as follows: the probability that we get a double ace is 1/36 so we have to throw the dice exactly 18 times to get a double ace at least once with probability 1/2. According to this reasoning, in more than 36 throws the probability that we get a double ace is more than 1, which is, of course, nonsense.

(*ii*) There are some "random quantities" which obey the "proportionality rule". (We shall discuss these random quantities in Paradox 8.) Some of these random quantities are very important in atomic physics, where the critical value is called half-life. This is inversely proportional to the decay constant, which corresponds to p.

(*iii*) The number of throws of a true dice necessary to turn up the first ace is a quantity depending on chance, a *random variable*. Let us denote this random variable by v. The possible values of v are $1, 2, 3, 4, \ldots$. The probability that $v=k$ (where k is a positive integer) is $\left(\frac{5}{6}\right)^{k-1}\frac{1}{6}$. So the *mean or expected value* of v, (defined as the weighted average of its possible values, the weights being the corresponding probabilities), is:

$$1\frac{1}{6}+2\frac{5}{6}\cdot\frac{1}{6}+3\left(\frac{5}{6}\right)^2\frac{1}{6}+4\left(\frac{5}{6}\right)^3\frac{1}{6}+\ldots=6.$$

Somewhat more generally, if p is the probability of the realization of an event A, and we repeat independent trials until A occurs, the expected value of the number of necessary trials is $1/p$. Thus these kinds of expected values obey the "proportionality rule": we need six times as many throws on average to get a double ace as to get one ace.

(*iv*) Up till now an intuitive term of probabilistical independence has been used. We shall return to this later.

(*v*) The explanation of de Méré's paradox did not become widely known. In 1693, nearly forty years after Pascal had solved the problem. *Samuel Pepys* (President of the Royal Society from 1684) proposed almost the same problem to Newton. Newton also found the right answer, but he could not satisfy Pepys either.

(*vi*) *"Dreydel"* or *"draydl"* is an ancient game similar to dice. (It also resembles the English game put-and-take.) Dreydel is played by Jews at the Chanukah festival. Quite recently Feinerman discovered (see reference below) that this game is unfair if the number of players is more than two, though, paradoxically, nobody had observed this fact for over 2000 years!

The *dreydel* is a four-sided top whose sides are denoted by the letters N, G, H and S (corresponding to the Hebrew letters Nun, Gimel, Hay and Shin). The game is played with any number of players, each of whom contributes one unit to the pot to start the game. The players continue to take turns spinning the *dreydel* until some mutually agreed stopping point. The payoffs (to the spinning player) corresponding to each of the four equally likely outcomes are, N: no payoff, H: half the pot, G: entire pot, S: put one unit into the pot. When one player spins a G, he collects the entire pot, and all the players then contribute one unit to form the new pot. If the number of players is denoted by m then the expected value of the payoff of the n-th spin is $E_n = m/4 + (5/8)^{n-1}(m-2)/8$.

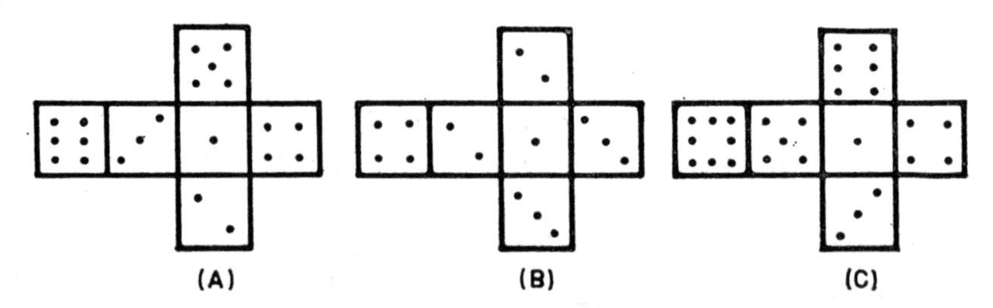

Figure 1. The faces of a regularly and two irregularly spotted dice. Rolling the two true dice which are irregularly marked by spots (B and C), the probability that the sum of the numbers we score is 2, 3, ..., 12 is the same as in the case of two regularly spotted dice.

We thus see that if $m>2$ then E_n is a strictly decreasing sequence. Therefore the first player (whose spins correspond to $n=1, m+1, 2m+1, \ldots$) has a term-for-term greater expected payoff than the second player......

e) References

Ore, Ø., "Pascal and the invention of probability theory", *The American Math. Monthly*, **67**, 409—419, (1960). This article summarizes the results of recent researches, concerning the role of de Méré and Pascal.

Rényi, A., *Letters on Probability*, Akadémiai Kiadó, Budapest, 1972. This book introduces to the early period of probability theory, in the form of four fictitious letters (from Pascal to Fermat).

Schnell, E. D., "Samual Pepys, Isaac Newton and the probability", *The American Statistician*, **14**, 27—30, (1960). This article is about the correspondence of Newton and Pepys.

Except Pascal's first letter (which disappeared) the correspondence between Pascal and Fermat was published in

Smith, D. E., *A Source Book in Mathematics*, McGraw-Hill, New York, 1929, 546—565 (new ed. 1959).

Feinerman, R., "An ancient unfair game", *The American Math. Monthly*, **83**, 623—625, (1976).

3. THE DIVISION PARADOX

a) The history of the paradox

This paradox was first published in Venice in 1494 in a summary of the mathematics of the Middle Ages. The author, *Fra Luca Paccioli* (1445—1509), entitled his book "Summa de arithmetica, geometria, proportioni et proportionalità". This book uses the word million and explains the rules of double entry for the first time. It is interesting to note that Fra Luca and *Leonardo da Vinci* became close friends in Milan and, due to this friendship, Leonardo illustrated Fra Luca's work "De Divina Proportione", which was published in Venice in 1509. Øystein Ore recently found an Italian manuscript dating from 1380 which also mentions the paradox of division. Many a thing indicates that the problem is of Arabic origin, or at least reached Italy through Arabic teaching.

However old this problem may be, it is a fact that it still took a very long time the question to be correctly solved. Paccioli himself did not even realize its connection with probability theory, for he considered it simply a problem in proportions. An incorrect solution was given by *Niccolo Tartaglia* (1499—1557), though he was such a genius that he discovered the formula to solve cubic equations in one night in a mathematical duel. After several unsuccessful attempts, Pascal and Fermat eventually gave the right answer to the problem independently of each other in 1654. It was such an important discovery that this date is considered by many people to be the birth of probability theory, and all the previous results to belong only to its prehistory.

b) The paradox

Two players are playing a fair game (i.e., both of them have the same chance of winning) and they have agreed that whoever wins 6 rounds first gets the whole prize. Let us suppose that the game actually stops before one of them wins the prize (e.g., the first player has won 5, the second 3 rounds). How could the prize be divided fairly? Though this problem is not, in fact, a paradox, the unsuccessful attempts of some of the greatest scientists to solve it, and the wrong, contradictory answers created the legend of a paradox. One of the answers was to divide the prize at the rate of the rounds won, i.e., 5:3. Tartaglia suggested a division at the rate of 2:1. (Most probably he thought that the first player had won two rounds more than the other, which is one third of the necessary 6 rounds, so the first player should get one third of the prize and the rest should be divided fifty-fifty.) As a matter of fact the fair rate is 7:1 which is far from the previous results.

c) The explanation of the paradox

Both Pascal and Fermat considered it a problem of probabilities. So the fair division is a rate of the chance of the first player to win against the second one. We shall calculate that in a case where the first player needs only one round to win while the second player needs three, the fair rate is 7:1. Following Fermat's idea we shall continue the game with 3 fictitious rounds even if some of them seem to be superfluous (i.e. when one of the players has already won the game). This extension makes all the possible $2 \cdot 2 \cdot 2 = 8$ outcomes equally probable. Since there is only one outcome when the second player gets the prize (i.e., when he wins all the 3 rounds) while in the other cases the first player wins, the fair rate is 7:1.

c) Remarks

(*i*) The general solution for the case when the first player needs n and the second player needs m more rounds to win was also due to Pascal and Fermat. The chance that the first player gets the prize is

$$\frac{1}{2^{n+m-1}} \sum_{j=n}^{n+m-1} \binom{n+m-1}{j}.$$

(Here the number of fictitious rounds is $n+m-1$ and all the possible 2^{n+m-1} outcomes are equally probable.) In 1654 the whole of Paris was talking about the discovery of a new science, i.e., the probability theory. Some months later a young genius, *Christian Huygens* arrived there from Holland to discuss with either Pascal or Fermat the problems of probability which he was also concerned with. As it happened he was unable to meet either of them. (Pascal was in ecstasy over religion and did not receive guests and Fermat lived far from Paris.) Nevertheless he heard of the most interesting results. He soon returned to Holland and began to write his book on probability theory. This excellent work, which also contains the solution of the problem of division for 3 players, was published in 1657 under the title of "De Ratiociniis in Aleae Ludo" as a part (the fifth book) of *Schooten's* "Exercitationes Mathematicarum". Huygens's work totals 16 pages and consists of a short preface and 14 propositions on gambling.

(*ii*) Fermat's beautiful idea of extending the play was applied by Anderson (see the reference below) in 1977! He reached the following striking theorem: Whether "service" is altered or the winner of one game serves next, the initial server will still have the same probabilities of winning N games before his opponent does.

e) References

Anderson, C. L., "Note on the advantage of first serve", *J. Combinatorical Theory* (A) **23**, 363, (1977).

Jordan, K., *Chapters on the Classical Calculus of Probabilities*, Akadémiai Kiadó, Budapest, 1972.

4. THE PARADOX OF INDEPENDENCE

a) The history of the paradox

First of all the independence of two random events A and B is defined. Let us denote their probabilities by $P(A)$ and $P(B)$ and let $P(AB)$ be the probability that both A and B occur. (The symbol P is widely used to denote the probability of an event, since not only in English but in many other languages the initial letter of the word "probability" is P — probabilitas in Latin, probabilité in French, probabilidad in Spanish, probabilità in Italian etc.) Let A be an arbitrary event and B an event with a positive probability. The probability of A, given that B has occurred, i.e., the conditional probability of A on the hypothesis B, will be denoted by $P(A|B)$ and defined by the ratio

$$P(A|B) = \frac{P(AB)}{P(B)}.$$

Two events A and B are said to be independent if equation

$$P(A|B) = P(A)$$

holds, that is, if the conditional probability equals the unconditional one. If we write the above equation in the form

$$P(AB) = P(A) \cdot P(B) \qquad (*)$$

we get a simple equation symmetric in A and B, where we do not even have to assume that $P(B)$ is positive. It is therefore preferable to start from the following definition: two events A and B are independent if equation (*) holds.

The mathematical definition of independence and what we generally think about independence are in harmony. For example, throwing two dice, the events "ace with first dice" and "ace with second dice" are clearly independent in an everyday sense and also in a mathematical sense. The harmony however did not seem to be perfect. It was *S. N. Bernstein* who called attention to the following paradox.

b) The paradox

When tossing two true coins, let A be the event "the first coin falls heads", B the event "the second coin falls heads" and C the event "one (and only one) of the coins falls heads". Then the events A, B and C are pairwise independent but any two of them uniquely determine the third one.

c) The explanation of the paradox

First of all it is obvious that A and B are independent since the result of the first throw is independent of the second one. The events A and C (and also B and C), however, do not seem to be independent at first sight, but, since $P(AC)=P(A)\cdot P(C)=\frac{1}{4}$ and similarly $P(BC)=$ $=P(B)\cdot P(C)$, they are really independent. It is also true that any two of the events determine the third one because each event (A, B and C) occurs exactly when one and only one of the other two events occurs. This paradoxical phenomenon shows that pairwise independence does not mean that events are independent as a whole. If we want to express the latter, we have to assume more than pairwise independence. A set of events is called mutually independent if for an arbitrary choice of finitely many events $A_1, A_2, \ldots, A_n$, the multiplication rule

$$P(A_1A_2 \ldots A_n) = P(A_1)\cdot P(A_2)\cdot \ldots P(A_n) \qquad (**)$$

holds, i.e., if the joint probability of the events is equal to the product of the individual (marginal) probabilities.

d) Remarks

(*i*) If the events $A_1, A_2, \ldots, A_n$ are not necessarily independent then we can only state that

$$-\left(1-\frac{1}{n}\right)^n \leqq P(A_1 A_2 \ldots A_n)-$$

$$-P(A_1)\cdot P(A_2)\cdot \ldots \cdot P(A_n) \leqq (n-1)n^{-n/(n-1)}.$$

(*ii*) Several simple paradoxes can be solved only with the help of the notion of independence. Let us examine the following problem. A boy is going to play three tennis matches against his mother and father, and he has to win twice in succession. The possible orders of matches are "father-mother-father" or "mother-father-mother". The boy has to decide which order is more favourable to him knowing that his father plays better than his mother. At first one might think that the second order is preferable to the boy as he plays twice with his mother in this version. Yes, but in this case the boy has to win the only match he plays against his father, otherwise he will not win twice in succession. Is it perhaps preferable to choose the first variation? If the boy wins against his father with probability p and with probability q against his mother, then $p<q$ since his father plays better than his mother. Choosing the first variation the boy has to win either the first and second matches, the probability of which is pq or the second and third ones, the probability of which is qp. Thus the probability that one of these two events will occur is $pq+qp-pqp$ (pqp has to be subtracted or else the probability of the boy winning three times is taken into account twice). Similarly if the boy chooses the second possible variation then the probability that he wins twice in succession is $qp+pq-qpq$. Since $p<q$, it follows that $pq+qp-pqp>qp+pq-qpq$, which means that it is preferable for the boy to choose the "father-mother-father" version!

(*iii*) We can also define the independence of random variables. Let $X_1, X_2, \ldots$ be arbitrary random variables assuming real values. The variables are called *mutually independent* (or independent, for short), if the events

$$A_{x_1} = \{X_1 < x_1\}, \quad A_{x_2} = \{X_2 < x_2\}, \ldots$$

are mutually independent for arbitrary real values of $x_1, x_2, \ldots$. The function $F(x)=P(X<x)$ is the *distribution function* of the random variable X, and the function $F(x, y, \ldots, w)=P(X<x, Y<y, \ldots, W<w)$ is called the *joint distribution function* for the random variables $X, Y, \ldots$ $\ldots, W$). Now we can define the mutual independence of any (finite of infinite) set of random variables in the following way: a set of random variables is called independent if for any finite subset S of this set, the joint distribution function of the random variables in S is equal to the product of their individual (marginal) distribution functions. If the distribution function $F(x)$ and the joint distribution function $F(x, y, \ldots w)$ can be written in the following form

$$F(x) = \int_{-\infty}^{x} f(\bar{x})\, d\bar{x}$$

and

$$F(x, y, \ldots, w) = \int_{-\infty}^{x} \int_{-\infty}^{y} \cdots \int_{-\infty}^{w} f(\bar{x}, \bar{y}, \ldots, \bar{w})\, d\bar{x} d\bar{y} \ldots d\bar{w},$$

then we call the functions $f(\bar{x})$ and $f(\bar{x}, \bar{y}, \ldots, \bar{w})$ *density functions*. If these density functions exist, independence means that the joint density function is equal to the product of the individual density functions.

(*iii*) If the density function $f(x)$ of the random variable X exists then its expectation is

$$E(X) = \int_{-\infty}^{\infty} x f(x)\, dx.$$

The expectation of $(X-E(X))^2$ is called the *variance* of X. Its positive square root is the standard deviation, which is a measure of the dispersion of X around its mean value. (There exist other measures of spread but standard deviation is undoubtedly the most important one. The first use of terms "standard deviation" and "variance" is due to K. Pearson (1895) and R. Fisher (1920), respectively.) If the density function of X is $f(x)$, then its variance

$$D^2(X) = \int_{-\infty}^{\infty} (X-E(X))^2 f(x)\, dx.$$

If X and Y are independent, then $E(XY)=E(X)E(Y)$ and $D^2(X+Y)=$ $=D^2(X)+D^2(Y)$ (provided that the variances of X and Y exist). The equation $E(X+Y)=E(X)+E(Y)$ holds without assuming the independence of X and Y.

e) References

Joffe, A., "On a set of almost deterministic k-dependent random variables", *Annals of Prob.* **2**, 161—162, (1974).

Wang, Y. H., "Dependent random variables with independent subsets", *The American Math. Monthly,* **86**, 290—292, (1979).

The following paper deals with a rather surprising relationship between independence and Peano curves

Holbrook, J. A. R., "Stochastic independence and space-filling curves", *The American Math. Monthly,* **88**, 426—432, (1981).

5. THE PARADOX OF BRIDGE AND LOTTERY

a) The history of the paradox

The history of games of chance can be traced back to ancient times. They became so widespread that certain states and religions considered it their duty to suppress them. *Frederick II*, emperor of the Holy Roman Empire, banned dice in 1232. (At that time it was probably the only popular game of chance.) *Louis IX*, King of France, decreed in 1255 that even dice making was illegal. In the Jewish Talmud a gambler was considered a thief and the Church pursued hazarders as well.

Among modern games of chance card games are undoubtedly the most widespread. The word "card" comes from the Greek word χαϱτης=paper, however card games *per se* date back to times preceding the invention of paper. Though we do not know where card games come from, they seem to have reached Europe through Venice via China—Persia—Syria—Palestine in the 13th century, at the time of the Crusades. The facts are as follows. According to a 17th century Chinese Encyclopedia, card type games were already known in China about 1120 AD. Parts of a 13th century Arabic card can be seen in the Istambul

Topkapi Serai Museum. A Florentine decree banned a card game called "naibi" in 1376. According to a 1377 manuscript, kept in the British Museum, card games became popular about that time in Switzerland. The Bibliothèque Nationale in Paris has 17 Tarot cards which were made for *Charles IV* in 1392. Johannes Gutenberg printed Tarot cards the same year as his famous Bible. The modern deck is derived from Gutenberg's Tarot cards. In the Tarot deck there were 78 cards; the 22 high-ranking cards were known as "atouts" (i.e., "above all"; later, these "atouts" were called "trump"). Some decades later the French dropped the 22 "atouts" and the 4 "Knights" the remaining 52 cards became the modern deck. Since then the number of the popular card games has risen to several hundred, while the number of the card sharpers has also increased. This fact is reflected in *Caravaggio's* famous painting "The Cardsharps" which was painted in 1593. In 1765 a Paris police lieutenant, *Gabriel de Sartine,* introduced roulette in order to reduce the influence of the sharpers. It became the most glamorous casino game and the oldest still in operation. Since the 17th century, lottery type games organized by the state have also become more and more popular. The first public lottery awarding money prizes, the Lotto de Firenze, was established in Florence in 1530. Another variation came into being in 1620 when the council of Genova needed five more members to be complete. These members were chosen from among 90 citizens whose names were put in an urn and the five names were pulled out. The citizens of Genova were allowed to bet on the five lucky citizens. Even today card games, roulette, lottery and other games of chance are very popular. Sometimes certain winning strategies appear claiming to be "absolutely reliable" but in fact they have no scientific background. On the other hand, exact scientific theories are only known by the very small society of mathematicians. These theories generally support the empirical rules used in practice. However, mathematical theorems may contradict common sense and become the source of the paradoxes. Here we only deal with two of them.

b) The paradox

(*i*) The paradox of bridge

Let us suppose that in a two-hand coalition of 26 cards there are 6 trumps altogether. Then the most probable distribution of trumps is the following: 4 in one hand and 2 in the other. The exact probability of this distribution is $\frac{78}{161}$, which is a little less than $\frac{1}{2}$, while the probability of the 3—3 distribution is just a little more than $\frac{1}{3}$, exactly $\frac{286}{805}$. Now suppose we have to throw trumps twice and in the coalition both hands can do so. In this case in the two-hand coalition there remain only 2 trumps. Either one hand has both cards or each of them has one. If 2 trumps and 20 other cards are distributed between two hands then the chance that one of them will get both trumps is $\frac{10}{21}$, while the probability of the other case is $\frac{11}{21}$. So the latter is more probable, i.e., the *more probable* distribution 1—1 comes from the *less probable* distribtuion 3—3. Is this not a contradiction?

(*ii*) The paradox of lottery

Most lottery players would not give a "too symmetrical" tip though every tip has the same chance. The reason is very simple. They know from experience that generally an irregular tip wins. In fact it is more advantageous to give a very symmetrical tip just because it is avoided by most of the other players.

c) The explanation of the paradoxes

(*i*) The chance that we get a 3—3 distribution of the trumps is:

$$\frac{\binom{6}{3}\binom{20}{10}}{\binom{26}{13}} = \frac{286}{805}.$$

Similarly for the 2—4 and 4—2 distribution it makes

$$\frac{\binom{6}{2}\binom{20}{11}}{\binom{26}{13}}+\frac{\binom{6}{4}\binom{20}{9}}{\binom{26}{13}}=\frac{78}{161},$$

so the second distribution is more probable. If we have 2 trumps and 20 other cards the 1—1 trump distribution has a chance of

$$\frac{\binom{2}{1}\binom{20}{10}}{\binom{22}{11}}=\frac{11}{21}.$$

The probability for the complementer event is of course $\frac{10}{21}$ as stated. But then where is the mistake? First of all we will show where it cannot be. It is natural to think that after trumping (and seeing that both hands have thrown 2 trumps) the probabilities have changed due to the information we have acquired meanwhile. It is true that the conditional probabilities (the condition is that both hands had at least 2 trumps) are different from those without condition but both probabilities are multiplied by the same number when we calculate the conditional probabilities. Consequently, their rate does not change and so the paradox is not solved. $\left(\frac{286}{805}+\frac{78}{105}=\frac{676}{805}\right.$ thus the conditional probabilities are $\frac{805}{676}$ times more than the ones without conditions.) The real source of error is the following. If the original distribution was 3—3 trumps, then both hands can throw their trumps in $3\cdot 2=6$ different ways. That gives $6\cdot 6=36$ possible events altogether. If the distribution was 4—2 or 2—4, then they could throw their trumps only in $4\cdot 3\cdot 2\cdot 1=24$ ways. As we can see now the original distribution of trumps before trumping is very important. If we take the original situation into consideration then we get a rate which is only $\frac{24}{36}=\frac{2}{3}$ of the rate calculated above. Really $\frac{286}{805}:\frac{78}{161}=\frac{11}{15}$ is $\frac{2}{3}$ of the rate $\frac{11}{21}:\frac{10}{21}=\frac{11}{10}$. Now the paradox is solved entirely.

(*ii*) It is not at all suprising that symmetrical or regular tips very seldom win. If the tip consists of 5 numbers then the possibilities by 90 numbers are about 44 million (exactly 43,949,268), while regular fives make only a few thousand. In the case of a regular 5-number-tip which is very seldom given by others, (though the chance of winning remains the same) the prize would actually increase. The player would certainly see it after a while if he played with a lot of lottery tickets.

d) Remarks

(*i*) Rarely given 5-number-tips can be traced back easily as it is always in the news how many players have got 2, 3, 4 or 5 hits and how much the prizes were. (If a more frequent 5-number-tip is drawn, prizes are less.) In case of football pools mathematical analysis is a bit more complicated because there are no fixed tips. Calculations may rely upon the tips of certain newspapers and the number of people taking their advice.

(*ii*) Works on games of chance (from roulette which is basicly hazard to bridge where the influence of randomness is reduced to minimum) must fill several libraries. In the 20th century the general theory of games was also developed mainly due to the work of *John von Neumann.* We will come to it later.

(*iii*) The following paradox appeared in 1693(!) in the *Philosophical Transaction of the Royal Society* (**17**, 677—681). "An Arithmetic Paradox, concerning the Chances of Lotteries" by the Honourable Francis Roberts, Esq; Fellow of the R. S.

"As some Truth (like the Axioms of Geometry and Metaphysics) are self-evident at the first View, so there are others no less certain in their Foundations, that have a very different Aspect, and without a strict and careful Examination rather seem repugnant. We may find Instances of this kind in most Sciences. ... I shall add one Instance in Arithmetic, which perhaps may seem as great a Paradox as any of the former.

There are two Lotteries, at either of which a Gamester paying a Shilling for a Lot or Throw; the First Lottery Upon a just Computation of the Odds has 3 to 1 of the Gamester, the Second Lottery but 2 to 1; nevertheless the Gamester has the very same disadvantage (and no more) in playing at the First Lottery as the Second."

The example following this problem points out (we use here modern terminology) as follows. Let X denote our prize depending on randomness in a game. (In case we lose X is negative.) Let $X^+=X$ if X is positive and otherwise 0, $X^-=X$ if X is negative and otherwise 0. Let Y, Y^+, Y^- denote the same random values in another game. Though the expected value of X and Y are the same according to the example the rate of the expected value of X^+ and X^- may differ from that of Y^+ and Y^-. This means that from $E(X)=E(Y)$ it does not follow that $E(X^+)/E(X^-)=E(Y^+)/E(Y^-)$. Hardly anybody would wonder about this result. (Obviously, if the expected value of both X and Y is 0 then the above rates are equal for both take the value 1.) If the problem is nevertheless considered a paradox, it should rather be called the paradox of expected value than "an arithmetic paradox".

e) *References*

Books on the mathematics of bridge are:

Borel, E.; Cheron, A., *Théorie mathématique du bridge*, Gauthier-Villars, Paris, 1940.

Jacoby, O., *How to Figure the Odds?* Doubleday, New York, 1947.

The author of the following article and book won millions in the game twenty-one and enforced the reformation of its rules.

Thorp, E. O., "A favourable strategy for twenty-one", *Proc. Nat. Acad. Sci.*, **47**, 110—112, (1961).

Thorp, E. O., *Beat the Dealer: A Winning Strategy of the Game Twenty-one* Blaisdell, New York, 1962.

For poker fans:

Findler, N. V., "Computer poker", *Sci. Amer.*, **239**, 112—119, (July 1978).

Works on the history of cards:

Beal, G., *Playing Cards and Their Story*, London, 1975.

Schreiber, W. L., *Die ältesten Spielkarten*, Stuttgart, 1937.

6. THE PARADOX OF GIVING PRESENTS; HORSE KICKINGS; TELEPHONE CALLS; MISPRINTS

a) The history of the paradox

Classical probability theory dealt mainly with combinatorial questions (connected with games of chance). In these problems random events usually had a finite number of possible outcomes and all outcomes had the same probability. In this simple case the probability of an event (A) is the ratio of the number of "favourable" cases to the "total number of cases". The first detailed monograph on probability theory also dealt with such probabilities. This was a book by *Rémond de Montmort* published in Paris 1708. The "Paradox of Giving Presents" is a variant of a problem discussed in Montmort's book in the language of card-games.

b) The paradox

The members of a company decide to give each other presents in the following way. Everybody brings a present, which is put with the others, mixed and distributed at random to the people. This is a fair way of distributing presents and is usually applied in the belief that the probability of a match, i.e., somebody getting his own present, is very small if the company is large. Paradoxically, the probability of at least one match is much larger than the probability of no matches (except if the company consists of exactly two members, when the chance of no matches is 50%).

c) The explanation of the paradox

Consider a company of n people; then the number of presents is also n. The presents can be distributed in $n!$ different ways. (This is the total number of cases.) The number of cases when nobody gets his own present is

$$\binom{n}{0} n! - \binom{n}{1} (n-1)! + \binom{n}{2} (n-2)! -$$

$$- \binom{n}{3} (n-3)! + \ldots + (-1)^n 0!,$$

thus the ratio of favourable cases to the total number of cases is:

$$p_n = \frac{1}{2!} - \frac{1}{3!} + \ldots + (-1)^n \frac{1}{n!},$$

and p_n is really smaller than $\frac{1}{2}$ if $n > 2$.

At the gathering of at least 6 people for example ($n \geqq 6$), $p_n \approx \frac{1}{e} \approx$ ≈ 0.3679 accurate to four decimal places. The probability of a certain match, i.e., that a certain person gets his own present, is clearly $\frac{1}{n}$, and $\frac{1}{n}$ converges to 0 as n increases. This paradox shows that "many a little makes a mickle": in spite of the small probabilites $\left(\frac{1}{n}\right)$ of certain matches, the probability of having at least one match is roughly 2/3.

d) Remarks

(*i*) The probability p_n converges to e^{-1} as n increases. If n is at least 6 then $p_n = e^{-1}$ accurate to four decimal places. More generally the probability of having exactly k matches is $\frac{e^{-1}}{k!}$ (in the above sense).

(*ii*) We shall examine another problem connected with the paradox of giving presents. Consider again a company of n people and n presents. Now the presents are distributed such that every person may get every present with the same probability independently of the distribution of other presents. Thus it may happen that somebody gets more than one present and others do not get any presents at all. Presents can be distributed now in n^n different ways (n^n is the total number of cases). Let A be the event that a certain person does not get any present. Then all the n presents are distributed among the remaining $(n-1)$ people and this can be done in $(n-1)^n$ different ways. Therefore the probability of event A is

$$q_n = \frac{(n-1)^n}{n^n} = \left(1 - \frac{1}{n}\right)^n.$$

The sequence q_n also tends to e^{-1} just like p_n did. Generalizing our result: the probability that a certain person gets exactly k presents converges to $\frac{e^{-1}}{k!}$ as $n \to \infty$. Consider now the case where the number of people (n) is not necessarily equal to the number of presents (m). In this case the probability we seek is $q_n = \left(1 - \frac{1}{n}\right)^m$. If the ratio $\frac{m}{n}$ tends to a parameter λ (i.e., if the average number of presents per person is λ or tends to λ) then q_n converges to $e^{-\lambda}$ (where λ can be an arbitrary positive real number). Finally the probability that a certain person gets exactly k presents converges to

$$r_k = \frac{\lambda^k e^{-\lambda}}{k!}.$$

We say that a random variable taking only non-negative integer values has the *Poisson distribution* if it assumes the value k with probability r_k.

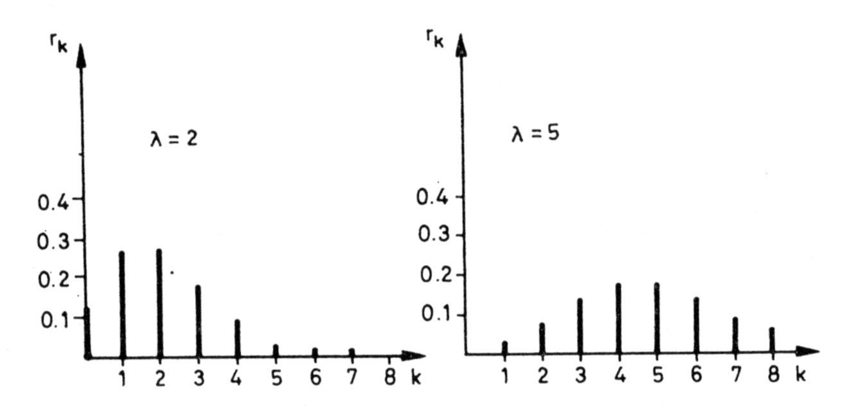

Figure 2. Poisson distribution with parameters $\lambda=2$ and $\lambda=5$.

As we have seen above the random number of presents that a certain person gets approximately follows the Poisson distribution with parameter λ, if the average number (expected value) of presents per person is λ. Returning to the "Paradox of Giving Presents" the number of people who get their own presents also follows the Poisson distribution with parameter $\lambda=1$, and this is quite natural since on the average there is only one person who gets his own present. (The probability that a certain person gets his own present is $\frac{1}{n}$ and for n people it adds to unity whatever the value of n.)

(*iii*) The notion of Poisson distribution appeared first in a book by the French scientist *Simeon Denis Poisson* (1781—1840). (Section 81 of "Recherces sur la Probabilité des Jugements en Matière Criminelle et en Matière Civile, Précédées des Règles Génerales du Calcul des Probabilités", published in 1837, deals with the application of probability theory in trials.) Poisson discussed the following problem. Consider an experiment in which the same phenomenon is repeatedly observed. It is assumed that the trials in this experiment are independent, there are only two possible outcomes for each trial and their probabilities remain the same throughout the trials. (An experiment of this type is called a sequence of Bernoulli trials.) It is usual to refer to the outcome with probability p as a "success" and the other as "failure". An example of Bernoulli trials is provided by successive tosses of an unbalanced coin. Let b_n be the probability that n (Bernoulli) trials result in k successes (e. g., the probability of exactly k heads in n tosses of an unfair coin). Then

$$b_k = \binom{n}{k} p^k (1-p)^{n-k}, \quad k = 0, 1, \ldots, n.$$

Thus the number of successes S_n in n Bernoulli trials is a random variable, which takes the value k with probability b_k. A random variable with possible values $0, 1, 2, \ldots, n$ is said to have binomial distribution if it assumes the value k with probability b_k. The attribute "binomial" refers to the fact that b_k is just the kth term of the binomial expansion of $(p+(1-p))^n$, since by the binomial formula $(p+(1-p))^n = b_0+b_1+$ $+b_2+\ldots+b_n$. Poisson discovered that if p is made smaller and smaller at the same time that n is made larger and larger, so that the product $np=\lambda$ is fixed, then b_k tends to r_k. Thus the Poisson distribution is an approximation to the binomial distribution. The wide applicability and great importance of the Poisson distribution was not realized in the middle of the last century; moreover it almost completely fell into oblivion. After 1894, however, it was applied to a very strange phenomenon. Statistics were made of how many soldiers had been killed by horse kicks during the 20 years between 1875 and 1894 in 14 different corps of the German Army. According to the 280 data, 196 soldiers had died this way, that is, $\lambda=0.7$ on average. If the number of fatal horse-kicks followed the Poisson distribution with the parameter $\lambda=0.7$, then we would expect no death in 139 cases, 1 death in 97 cases, 2 deaths in 34 cases, etc. out of the 280 cases. And what did the statistics show? The actual data were 140, 91, 32 etc., respectively; practice and theory are in such close agreement that we would have hardly expected more.

This comparison appeared in 1898 in the famous monograph by *L. Bortkiewicz.* The title of his book "Law of Small Numbers" refers to the fact that in the Poisson approximation p tends to 0 as n increases. (The title is quite misleading since it suggests that the Poisson approximation is in contrast in some way with the laws of large numbers, which will be discussed later on). The Poisson distribution only began to be applied widely in the 20th century. For example, the number of certain kinds of goods sold on a given day approximately follows the Poisson distribution, or the number of hemoglobins visible under the microscope, the number of strikes and wars in a year, the number of misprints in a text, or the number of telephone connections to a certain number on a certain day also follow approximately the Poisson distribution. If the average number of hemoglobins or misprints or telephone connections is λ then they follow the Poisson distribution with parameter λ. If N telephone lines are available in a telephone exchange then the number of busy lines has approximately the Poisson distribution. Problems of this type were studied by the Danish mathematician *A. K. Erlang* (1878—1929). He pointed out, in 1906, that a better approximation can be obtained by using the following truncated Poisson distribution:

$$e_k = \frac{c\lambda^k}{k!}, \quad \text{where} \quad k = 0, 1, \dots, N,$$

and

$$c = \left(\sum_{k=0}^{N} \frac{\lambda^k}{k!}\right)^{-1}.$$

Since that time this has been called the Erlang distribution.

e) References

Haight, F. A., *Handbook of the Poisson Distribution,* Wiley, New York, 1967.
Montmort, P. R., *Essay d'analyse sur les jeux de hazard,* Paris, 1708.
Sheynin, O. B., "S. D. Poisson's work in probability", *Archive for the History of Exact Sciences,* **18**, 245—300, (1978).
Stigler, S. M., "Poisson on the Poisson distribution", *Statistics and Probability Letters,* **1**, 33—35, (1982).
Takács, L., "The problem of coincidences", *Archive for the History of Exact Sciences,* **21**, 229—245, (1980).

7. ST. PETERSBURG PARADOX

a) The history of the paradox

Probability theory, which originally described exepriences connected with games of chance, developed into a theory of great universality and gained ground in many fields of life. Thus it was not surprising that almost every notable scientific journal followed the example of the English "Philosophical Transactions" and published articles on probability theory regularly. More and more scientists thought that probability was none other than the very guide of life, reason in terms of figures. However, in the early 1700s the Academy of St. Petersburg published an article in which the mathematical calculation did not seem to be in harmony with reason. *Daniel Bernoulli* wrote the article and made the Petersburg paradox known, but it was his cousin *Nicolaus Bernoulli* who had first raised the problem and mentioned the paradox in a letter written to Montmort in September 1713. (The Bernoullis were a renowned family of mathematicians, several members of which dealt with probability theory, especially *James Bernoulli,* who will be mentioned later in connection with the laws of large numbers.)

b) The paradox

A single trial in the Petersburg game consists of tossing a fair coin until it falls heads; if this occurs at the rth throw the player receives 2^r dollars from the bank. Thus the gain redoubles at each toss. The question is the following: how much money should the player pay as an entrance fee so that the game will become fair? The Petersburg game was considered fair in the classical sense, if the mean (or expected) value of the net profit is 0, but surprisingly we cannot fulfill this natural requirement no matter how much (finite) money we pay.

c) The explanation of the paradox

The loss of the bank has infinite expectation since the probability that the game ends at the kth toss is $1/2^k$ and in this case the player receives 2^k dollars. Then the bank has to pay

$$\frac{1}{2}\cdot 2+\frac{1}{4}\cdot 4+\frac{1}{8}\cdot 8+\ldots = 1+1+1+\ldots$$

dollars on average which is an infinite quantity of money, so an infinite amount of money would be a fair entrance fee. Though this calculation is mathematically correct, the result was unacceptable, therefore several mathematicians suggested more acceptable modifications.

(*i*) *Buffon, Cramer* and others suggested accepting the natural assumption of limited resources (i.e., only limited amount of money available for the bank). Let this amount of money be one million dollars. Then the expected value of the player's gain is

$$\frac{1}{2}\cdot 2+\frac{1}{4}\cdot 4+\ldots+\frac{1}{2^{19}}\cdot 2^{19}+$$

$$+\left(\frac{1}{2^{20}}+\frac{1}{2^{21}}+\ldots\right)10^6 = 19+1.90\ldots \approx 21,$$

(taking into consideration that $2^{20}>10^6$). Therefore if the player pays a 21 dollar entrance fee then the game becomes somewhat favourable to the bank.

(*ii*) *W. Feller* pointed out that it is possible to determine entrance fees which would make the Petersburg game fair. Denoting by n the number of games the player played, the game can be considered fair if the ratio of the accumulated gain N_n to the accumulated entrance fees R_n converges to 1 as n tends to infinity, more precisely if for every $\varepsilon > 0$

$$P\left\{\left|\frac{N_n}{R_n} - 1\right| < \varepsilon\right\} \to 1 \quad \text{as} \quad n \to \infty. \tag{*}$$

Feller proved that the Petersburg game becomes fair if we put $R_n = n \cdot \log_2 n$. By the paradox the game cannot be fair for $R_n = cn$, where c is any finite constant. If, however, entrance fees may depend on the number of games the player played then (according to Feller's theorem) the Petersburg paradox vanishes.

d) Remarks

(*i*) The relation (*) expresses a stability property of N_n. Similar stabilities will appear with $R_n = cn$ in "The paradox of Bernoulli's law of large numbers".

(*ii*) On the results of 2084 games, *Buffon* found that the game becomes fair with about 10 dollars entrance fee.

(*iii*) The following paradox is a companion to the St. Petersburg paradox. (I heard it from Sam Gutmann after my talk in Dudley's seminar at MIT in 1983.) Say you are given an opportunity to win $(-2)^n$ dollars with probability 2^{-n}, $n = 1, 2, 3, \ldots$ Are you happy or sad? The answer is you are both happy *and* sad. You are happy because the given lottery is equivalent to a compound lottery in which you receive one of a list of lotteries, each of which is favorable (has positive expectation). That is, you receive with probability $(2^{-1} + 2^{-2} + 2^{-4})$ the lottery which awards you $(-2)^j$ dollars with probability

$$\frac{2^{-j}}{2^{-1} + 2^{-2} + 2^{-4}} \quad (j = 1, 2, 4)$$

or, with probability $(2^{-3} + 2^{-6} + 2^{-8})$ you receive the lottery which awards you $(-2)^k$ dollars with probability

$$\frac{2^{-k}}{2^{-3} + 2^{-6} + 2^{-8}} \quad (k = 3, 6, 8),$$

etc. Each of these individual three-award lotteries has positive expectation. So you are happy. Of course you can also rewrite the original lottery into three-award lotteries each of which has negative expectation. The first has rewards $(-2)^1, (-2)^2, (-2)^3$, the second has rewards $(-2)^4$, $(-2)^5, (-2)^7$, etc. So you are sad, too. [Here is a restatement for those who are familiar with the notion of conditional expectation. Imagine that the conditional expectation $E(X|Y)$ is defined, not as usual, but rather as $\int xP(dx|Y)$ where $P(dx|Y)$ is defined as usual. Then there exist random variables X, Y, and Z such that $E(X|Y)>0>E(X|Z)$ with probability 1! Simply let X be the ultimate reward in the lottery, i.e., $X=(-2)^n$ with probability 2^{-n}. Let $Y=1$ if we receive the first positive lottery (i.e., $X=(-2), (-2)^2$, or $(-2)^4$), let $Y=2$ if we receive the second (i.e., $X=(-2)^3, (-2)^6$, or $(-2)^8$), etc. Let $Z=1$ if we receive the first negative lottery (i.e., $X=(-2), (-2)^2, (-2)^3$), $Z=2$ if we receive the second (i.e., $X=(-2)^4, (-2)^5, (-2)^7$) etc.]

e) References

Feller, W., *An Introduction to Probability Theory and Its Applications*, New York, John Wiley, 1969.

Hinčin, A. Ja., "Su una legge dei grandi numeri generalizzata", *Giorn. Ist. Ital. d. Attuari*, 7, 365—377, (1936).

Martin—Löf, A. "A limit theorem which clarifies the 'Petersburg paradox'" *J. Appl. Prob.* **22**, 634—643, (1985).

8. THE PARADOX OF HUMAN MORTALITY. THE AGELESS WORLD OF ATOMS AND WORDS

a) The history of the paradox

The mathematical research on human mortality and life span began in the early days of capitalism due to the demands of the insurance companies. Following the results obtained by *John Graunt* (1662), *van Hudden* and *John de Witt* (1671) in the 1660s, *Edmond Halley* (the discoverer of the comet named after him) published a paper in 1693 on mortality tables establishing the mathematical theory of life insurance. The following paradox (raised by d'Alembert) shows one of the "teething troubles" of the new theory.

b) The paradox

In Halley's table the average life span is 26 years, and yet one still has an equal chance of dying before the age of 8 or living beyond the age of 8.

c) The explanation of the paradox

It is true that according to Halley's table one has an equal chance of surviving more than 8 years and dying before 8, but once he has already lived for 8 years he can still live for several decades. Therefore it is not surprising that the average life span is much more than 8. Supposing that out of a thousand people only one attains the life span of Methuselah, the average age will increase a lot but their probable age (which they survive at a chance of 50%) will not change significantly.

d) Remarks

(*i*) Let $F(x)$ denote the probability that in a population the life span of a randomly choosen person is less than x time units. ($F(x)$ is the distribution function of life span.) Suppose that $F(x)$ has a density function $f(x)$. The average life span is $M=\int_0^\infty xf(x)dx$. On the other hand the probable life span m is defined by the equation $F(m)=\frac{1}{2}$. In other words, during the time period m half of the population dies out. It is clear from these formulas that generally M and m are of quite different values. While M is the expected value of life span, m is called its median.

(*ii*) The notion of human mortality can easily be extended. If we consider the amortization of industrial products or the decay of the atoms as death then we obtain a widely applicable mathematical theory developed from the study of human mortality. However, in this more extended field rather paradoxical phenomena may arise as well. While human beings are neither immortal nor ageless, we can find ageless beings both in nature and society. Let us define the notion of agelessness. Consider a being ageless if the chance that it will survive a certain fixed time interval is independent of the time it has already "lived". Naturally, man does not possess this feature for the longer he lives the more probably he will die in a given period of time. It is interesting that not every being imitates us. For example radioactive atoms are ageless beings. If the average life span of an ageless being is T then the probability that it will not die in the following time period x is $e^{-x/T}$, where x is a positive number. The ageless property of radioactive particles follows from the fact that their speed of decay is in proportion to the number of undecayed particles. The factor of proportionality is called the decay constant and is denoted by λ. If there were N_0 undecayed particles at the moment $t=0$ then (as the speed of decay is constant we get by integration that) at the moment x the number of the undecayed particles is $N_x=N_0e^{-\lambda x}$. It means that the survival probability for the moment x is $e^{-\lambda x}$. Consequently, the radioactive particles are really ageless and their average life span is $T=\frac{1}{\lambda}$. In other words, the life span of radioactive particles follows an *exponential distribution* with parameter λ, i.e., its density function is $\lambda e^{-\lambda x}$. The half-life of ageless beings (the period during which half of the beings die out) is the root of the following equation:

$$e^{-\lambda x}=\frac{1}{2}, \quad \text{namely} \quad x=\frac{\ln 2}{\lambda}.$$

(*iii*) The half-life of ageless beings has become a fundamental idea in several fields of science. The radiocarbon method, worked out by the American chemist *Willard Frank Libby,* is still the most applied dating method in the field of archeological chronology. (The scientist was awarded the Nobel Prize for this discovery in 1960.) In 1950, following Libby's ideas, *M. Swadesh* applied his method to linguistics assuming that not only radioactive atoms but lingual atoms, i.e., words can also be considered ageless. The ancient basic vocabulary of languages dies out at a supposed half-life of 2000 years. With the help of this idea we can determine the date when two related languages (e.g., Latin and Sanskrit) separated. We only have to know the amount of basic vocabulary still existing in both languages to be able to figure out the date they separated. *A. Raun* and *E. Kangsmaa-Minn* compared Hungarian and Finnish. They found that the identical elements make 21% and 27%, resp. (The calculations were made by other methods.) On the basis of this, Hungarian and

Finnish are thought to have separated some 4—5000 years ago. Swadesh's 30-year-old method is very often used and is known as lexicostatistics or glottochronology. (Swadesh's original article was published in the International Journal of American Linguistics.)

(*iv*) Suppose the decay constant is λ. Then the probability that exactly k particles will decay in a time period t is

$$\frac{(\lambda t)^k e^{-\lambda t}}{k!}.$$

This means that the number of decays is a Poissonian random variable already known from the paradox of giving presents. The expected value of this distribution is λt, which is quite natural.

(*v*) We have seen that there exist ageless beings. What is even more surprising is the existence of beings growing younger, e.g., machines during their running in time, when the probability that they will not go wrong for a certain period increases with the passing time. It can easily be seen that mathematically this means that, using the notations of (*i*), $\frac{f(x)}{1-F(x)}$ (the failure rate) is a decreasing function of x. The examination of this rate is fundamentally important in reliability theory and in storage problems.

(*vi*) Finally we mention a fascinating question related to human mortality. Can the total number of people who ever lived on the Earth be estimated by some probabilistic methods? The background of the following surprising statement is explained in Goldberg's book (see below). "9 percent of everyone who ever lived is alive now." This sentence was also the title of an article in *The New York Times* (Oct. 6, 1981. p. 61).

e) References

Barlow, E. R., Proschan, F., *Statistical Theory of Reliability and Life Testing*, New York, Holt, Rinehart and Winston, Inc., 1975.

Belyaev, Yu. K., Gnedenko, B. V., Solovev, A. B., *Mathematical Methods in the Theory of Reliability*, Moscow, Nauka, 1965. (in Russian)

Goldberg, S., *Probability in Social Science*, Birkhäuser, Boston—Basel—Stuttgart, 1983.

Halley, E. "An estimate of the degrees of mortality of mankind, drawn from curious tables of the births and funerals at the city of Breslau; with an attempt to ascertain the price of annuities upon lives", *Philosophical Transactions of the Roy. Soc.*, **17**, 596—610, 654—656, (1963).
(It is interesting that Halley though being English did not use the data of London or Dublin but that of Breslau (today it is Wroclaw), which was the capital of Silesia at that time. He used the monthly birth and mortal data of the period 1687—1691 of that area because he thought that the disturbing effects of migration were much less in Breslau.)

9. THE PARADOX OF BERNOULLI'S LAW OF LARGE NUMBERS

a) *The history of the paradox*

There are only few other laws in mathematics that have been as much misunderstood as the laws of large numbers. (It is not even generally known that several such laws exist.) The first law of large number was proved by *Jacob Bernoulli* (1654—1705) in his book entitled "Ars conjectandi" (Art of Guessing) which was published only after his death in 1713. Bernoulli himself did not use the notion "law of large numbers", it was introduced only by Poisson in 1837. According to Bernoulli's law, if we toss a fair coin n times and it falls k times heads, then, by increasing the number of tosses (n), the rate k/n (the relative frequency for tossing heads) will approach the value 1/2. More precisely if ε and δ are arbitrary small positive numbers and n (depending on ε and δ) is great enough then $|k/n-1/2|$ is less than ε with a probability of at least $1-\delta$. This theorem is not nearly as complicated as one might think from the number of misunderstandings and paradoxes it caused. The most typical is as follows.

b) *The paradox*

Gamblers often believe that, according to the law of large numbers, if a fair coin falls heads many times then the probability of tossing tails will necessarily increase. (Otherwise it would not be true that after a great many tosses the number of heads and tails are approximately the same.) On the other hand, it is obvious that coins cannot remember and so they do not know how many times they have already fallen tails or heads. For this reason in every toss the chance of heads is 1/2 even if the coin has already fallen heads a thousand times in a row. Is this not in contradiction with Bernoulli's law?

c) The explanation of the paradox

According to Bernoulli's law, the number of heads and tails must be approximately equal in the case of a great many tosses, but here the point is what is meant by "approximately". The gambler who believes that the *difference* between the number of heads and tails must be very small is mistaken for Bernoulli's law only states that the *rate* of the number of heads and the total number of tosses is approximately $\frac{1}{2}$ (with a probability close to 1) or equivalently, the rate of the number of heads and tails approximates 1, in other words, the *difference of their logarithms* approximates 0 (provided that the number of tosses increases). If the difference itself should remain small, it would contradict the lack of memory property of coins.

d) Remarks

(*i*) It is clear now that no matter how many times we observe heads successively at the next toss the chance of tails will by no means be greater. The following question now arises. Let us suppose that we toss a coin n times. What is the longest run of heads only we can expect? Tossing a coin n times if $n=100$ then we can expect 6—7 heads successively, if $n=1000$ then we can expect 9—10, and 19—20 for $n=10^6$. The following theorem was proved by *Paul Erdős* and *Alfréd Rényi*. If we toss a coin n times then there occurs a "pure head" run of length $\log_2 n$ with a probability converging to 1 as $n \to \infty$. This fact is very useful in deciding whether a sequence of two signs describes the result of coin tosses or somebody has created it "carefully" avoiding long runs. Owing to the ingrained misunderstanding of Bernoulli's law of large numbers, most people would not write the same sign consecutively 7 or more times in a sequence of 100 signs.

(*ii*) According to the above remark pure runs (heads or tails) may be rather long. On the other hand, it is easy to calculate that the expected length of the 1st, 2nd, 3rd, etc. pure runs is always equal to 2. If your coin is not fair and the probability of tossing heads is $0<p<1$ then the expected length of the 1st, 3rd, ... (every odd) pure run is $\frac{p}{q}+\frac{q}{p}$, while the expected length of the 2nd, 4th, ... (every even pure run) is always 2 (independently of p(!)). The sum $\frac{p}{q}+\frac{q}{p}$ cannot be less than 2, which means that the odd runs are at least as long as the even runs. This fact is by no means surprising because it is more probable for a coin to fall on its more probable side first. Thus the first run has a greater chance for being long than short, so on the average it is long or at least longer than the second run where the expected length is only 2. What is surprising is its independence of p.

(*iii*) Bernoulli's law of large numbers can be expressed concisely by the help of the notion of *convergence in probability*. We say that a series of random variables $X_1, X_2, \ldots$ converges in probability to a random variable X, if the probability of $|X_n-X|>\varepsilon$ converges to 0 for every positive ε, i.e., if $P(|X_n-X|>\varepsilon)\to 0$. (Paradoxically, it may occur that a series of random variables $X_1, X_2, \ldots$ converges in probability to 0 but

$$\frac{X_1+X_2+\ldots+X_n}{n}$$

does not.) Bernoulli's law says that the relative frequency k/n of an event converges in probability to its probability. To prove convergence in probability we generally use the Čebyshev—Bienaymé inequality, according to which if the expected value of X is E and its variance is D^2 then

$$P(|X-E|>\varepsilon) \leqq \frac{D^2}{\varepsilon^2}.$$

It is interesting that the Russian P. L. Čebyshev and the French J. Bienaymé published their inequality which they had discovered independently in the very same number of a French journal. (*J. Math. Pures et. Appl.* XII. 1867). From this inequality it follows at once that if the distribution of the independent random variables $X_1, X_2, \ldots$ is the same and their variance D^2 is finite then the arithmetical mean

$$\frac{X_1+X_2+\ldots+X_n}{n}$$

converges in probability to the common expected value of X_i (the variance of this arithmetical mean is $\frac{D^2}{n}$, which converges to 0 if $n\to\infty$). This is one of the general (weak) laws of large numbers. The weak laws of large numbers examine convergence in probability while strong laws describe convergence with probability 1. The next remark concerns the latter.

(*iv*) Among strong laws of large numbers the best known is Kolmogorov's theorem: if $X_1, X_2, \ldots$ are mutually independent random variales with the same probability distribution function having a (common) finite expected value E then the arithmetical mean

$$\frac{X_1+X_2+\ldots+X_n}{n}$$

converges to E as $n\to\infty$ with probability 1. If the random variables are positive and $S_n^{(k)}$ denotes the elementary symmetric polynomial

$$\sum_{1\leq j_1<j_2\ldots<j_k\leq n} X_{j_1}X_{j_2}\ldots X_{j_k}$$

then

$$\sqrt[k]{S_n^{(k)}\Big/\binom{n}{k}}$$

also converges to E (with probability 1) provided $k=k(n)$ is a natural number so that $k/n\to 0$ as $n\to\infty$. The limit exists with probability 1 even if k/n tends to a positive number c. Then this limit is a constant depending on c (if $0<c<1$ then it is enough to suppose that the expected value of $\log(1+X_i)$ is finite, and the same holds for $|\log X_i|$ if $c=1$; see the paper by Halász and Székely). If the random variables can take both positive and negative values then the problem is more complicated.

(*v*) The "law of large numbers" is false in the sense of category (see the paper by Méndez).

e) References

Erdős, P., Rényi, A., "On a new law of large numbers", *Jour. Analyse Mathematique,* **23**, 103—111, (1970).
Erdős, P., Révész, P., "On the length of the longest head-run", *Coll. Math. Soc. J. Bolyai,* **16**, (ed. Csiszár, I., Elias, P.) (1975).
Halász, G., Székely, G. J., "On the elementary symmetric polynomials of independent random variables", *Acta Math. Acad. Sci. Hung.,* **28**, 397—400, (1976).
Komlós, J. A generalization of a problem of Steinhaus. *Acta Math. Acad. Sci. Hung.,* **18**, 217—229, 1967.
Méndez, C. G. On the law of large numbers, infinite games, and category. *The American Math. Monthly,* **88**, 40—41, 1981.
Móri, T. F., Székely, G. J. Asymptotic behaviour of symmetric polynomial statistics. *Annals of Probability,* **10**, 124—131, 1982.
Móri, T. F., Székely, G. J. Asymptotic independence of "pure-head" stopping times. *Statistics and Probability Letters,* **2**, 5—8, 1984.
Révész, P. The Laws of Large Numbers. Akadémiai Kiadó, Budapest, 1967.

10. DE MOIVRE'S PARADOX; ENERGY SAVING

a) The history of the paradox

One of the most outstanding figures of probability theory is *Abraham de Moivre* (1667—1754). He was a mathematician of French origin but after the revocation of the Edict of Nantes (which provided the Huguenot's freedom of religion), he moved to England. His fundamental work "The Doctrine of Chances" was published there in 1718. In the third edition of the book (1756), de Moivre himself writes enthusiastically about his epoch-making discovery (already communicated to some of his friends in 1733), which proves much more than Bernoulli's law of large numbers: "... I'll take the liberty to say, that this is the hardest problem that can be posed on the subject of Chance...". There is no doubt that de Moivre's discovery, the normal distribution, has become a pillar of the science of chance. (Curiously enough, de Moivre did not incorporate it in the second edition (1738).)

b) The paradox

According to Bernoulli's law of large numbers, in the coin tossing game the probability that the number of heads the player scores is approximately equal to the number of tails tends to 1 as the number of tosses increases (approximate equality means that the ratio of the two numbers tends to 1). On the other hand, the probability that the number of heads is exactly equal to the number of tails tends to zero. For example, in 6 tosses of a coin the probability of scoring 3 heads is 5/16; in 100 tosses the probability of scoring 50 heads is 8%; in 1000 tosses the probability of scoring 500 heads is less than 2%. Generally, when tossing $2n$ times, the the probability that it falls heads exactly n times is $p=\binom{2n}{n}\Big/2^{2n}$ and, for sufficiently large n, p is approximately $1/\sqrt{\pi n}$, which really tends to zero as n increases. In sum: the probability that the number of heads approximately equals the number of tails tends to one, whereas the probability that the number of heads is exactly equal to the number of tails tends to zero. The gulf between the two facts was surrounded by a "paradoxical atmosphere" till de Moivre succeeded in building a mathematical bridge over the gulf.

c) The explanation of the paradox

Let H_n and T_n denote the number of heads and tails, respectively, in n tosses of a coin. According to Bernoulli's law of large numbers, the probability that H_n-T_n becomes negligibly small compared to n tends to one (what is not at all surprising). De Moivre, however, noticed that the term $|H_n-T_n|$ is not negligible compared to $\sqrt{n}$. He calculated, for example, for $n=3600$ that the probability that $|H_n-T_n|$ is at most 60 is 0.682688.... Let x be an arbitrary positive number and let $A_n(x)$ denote the probability that $|H_n-T_n|<x\sqrt{n}$. According to de Moivre, $A_n(x)$ tends to a value $A(x)$ which is between 0 and 1 as n increases. When x begins to increase from zero to infinity, $A(x)$ also increases steadily from zero to one (see Remark (*i*)). This function $A(x)$ is the above-mentioned bridge over the gulf. To determine $A(x)$, de Moivre used Stirling's

formula, which he had also discovered independently of Stirling. (*James Stirling's* formula was proved in 1730 and states that $n!$ is asymptotically equal to $\sqrt{2\pi n}\left(\frac{n}{e}\right)^n$.)

d) Remarks

(*i*) The exact form of the function $A(x)$ is:

$$A(x) = \sqrt{\frac{2}{\pi}} \int_0^x e^{-u^2/2}\, du.$$

Using this formula, de Moivre's theorem can be written is the following form:

$$\lim_{n\to\infty} P(|H_n - T_n| < x\sqrt{n}) = A(x), \quad \text{if} \quad x > 0,$$

or

$$\lim_{n\to\infty} P(H_n - T_n < x\sqrt{n}) = \Phi(x),$$

where

$$\Phi(x) = \frac{1}{\sqrt{2\pi}} \int_{\infty}^{x} e^{-u^2/2}\, du.$$

A random variable which assumes values smaller than x with probability $\Phi(x)$ (where x is an arbitrary real value) is said to obey a *standard normal distribution.* According to de Moivre's result, $(H_n - T_n)/\sqrt{n}$ approximately obeys a standard normal distribution (if n is large enough).

[Table 1 at the back of the book gives the values of $\Phi(x)$.]

(*ii*) Since $H_n + T_n = n$, the above result can be reformulated in the following way:

$$\lim_{n\to\infty} \left(H_n < \frac{n}{2} + x\frac{\sqrt{n}}{2}\right) = \Phi(x).$$

De Moivre also examined the case where the coin was a biased one (not fair) and it fell heads with probability p, and fell tails with probability $1-p$. Then

$$\lim_{n\to\infty} P(H_n < np + x\sqrt{np(1-p)}) = \Phi(x),$$

which is known as the "*de Moivre—Laplace limit theorem*". This theorem can be widely applied in a whole range of plannings, e.g., energetics.

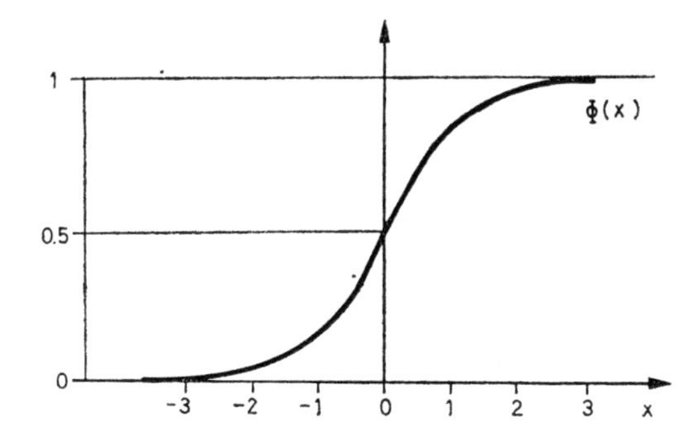

Figure 3. The standard normal distribution function.

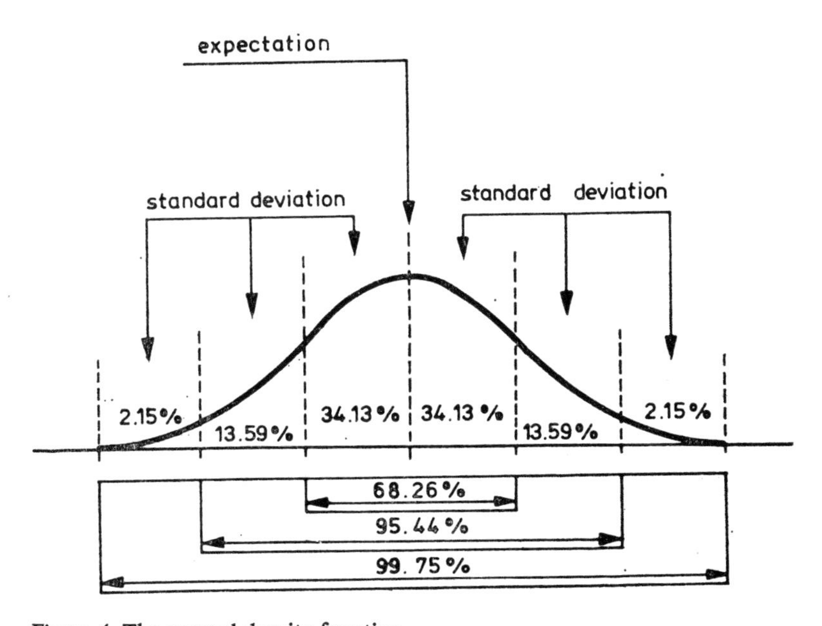

Figure 4. The normal density function.

Example: Consider 300 similar machines in a factory. If, on average, 70% of the machines work and 30% of them are under repair then power has to be provided for 210 machines, on average. Sometimes, however, all the 300 machines may be working. How much power has to be provided to be 99.9% sure that every machine will have enough power to work? (It is assumed that every good machines go wrong independently of each other.) In the above formula, H_n now stands for the number of working machines, $n=300$ and $p=0.7$. According to Table 1, $\Phi(x) \approx \approx 0.999$ if $x=3$. Using these values, $np+x\sqrt{np(1-p)}=210+3\sqrt{63}$, so it is enough to take into account 234 machines. (In practice, however, nearly all the 300 machines are taken into account, being unnecessarily overcautious.)

(*iii*) The de Moivre—Laplace theorem, discussed above, can be generalized in many ways. The members of the St. Petersburg mathematical school, led by *P. L. Čebyshev* (1821—1894), especially *A. M. Liapunov* (1857—1918) and *A. A. Markov* (1856—1922), gained great distinction for generalizing the de Moivre—Laplace theorem. Let $X_1, X_2, \ldots$ be mutually independent random variables with a common distribution (i.e., with a common distribution function). Suppose that the expectation and the standard deviation of the random variables exist and are finite.

Let M denote the common expectation and D the standard deviation of the random variables $X_1, X_2, \ldots$ and let $S_n = X_1 + X_2 + \ldots + X_n$. Then

$$\lim_{n \to \infty} P\left(S_n < nM + xD\sqrt{n}\right) = \Phi(x).$$

This is the *central limit theorem,* the most important of all limit theorems (it is because of its importance that it is called "central", a denomination first used by George Pólya). In general, limit theorems discuss the asymptotic distributions of different functions (e.g., the sum, product, maximum, etc.) of random components. The central limit theorem — and its generalizations — explains why we meet the normal distribution in nature so often, especially in connection with quantities which can be composed from many ("nearly") identically distributed ("nearly") independent random components. However, it is worth emphasizing that the "composition" of random variables in nature is not always their sum, so the investigation of the behaviour of other functions of random variables is very important. The known limit theorems do not explain completely the frequent occurrence of the normal distribution. According to *Poincaré's* sarcastic remark, everybody believes in the universality of normal distribution: physicists believe in it because they think that mathematicians have proved its logical necessity, and mathematicians believe in it because they think that physicists have verified it by laboratory experiments.

e) References

Gnedenko, B. V., Kolmogorov, A. N., *Limit Distributions of Sums of Independent Random Variables* (in Russian), Gostehizdat, M.-L., 1949. English edition: Reading, Mass.: Addison-Wesley, 1954.

Ostrowski, A., "On the remainder term of the de Moivre—Laplace formula (To the 70th birthday of Eugene Lukacs)", *Aequa. Math.*, **20**, 263—277, (1980).

Petrov, V. V., *Sums of Independent Random Variables,* Academie-Verlag, Berlin, 1975.

Székely, G. J., "A limit theorem for elementary symmetric polynomials of independent random variables", *Z. Wahrscheinlichkeitstheorie verw. Geb.*, **59**, 355—359, (1982).

11. BERTRAND'S PARADOX

a) The history of the paradox

Georges Buffon (1707—1788), the famous French scientist, founded a new branch of probability theory with a paper written in 1733 (but published in 1777). The solution of the celebrated "needle problem" discussed in this paper required a geometrical (rather than combinatorical) method. In these sort of problems the random points considered are supposed to be uniformly distributed in a given domain. (E.g., the bullets on a score-card.) The probability of falling into any part of a given domain is in proportion to its area (length or volume). Thus to calculate the probability we only have to compute the quotient of the "favourable" and the "total" area (length or volume). These kinds of probabilities also resulted in several paradoxes. E.g., the chance of hitting the very middle (or any other fixed point) of a score-card is obviously 0. On the other hand, it is not impossible to hit this point and therefore we must distinguish an event of probability 0 from the impossible event (the probability of the impossible event is 0 but the opposite is not true).

It also sounds very strange that both hitting at least one of finitely many points and hitting only one of them have the same probability. (Both probabilities are equal to 0. See the paradox of zero probability.) Another curiosity: a one-to-one transformation may completely change the chances. E.g., if we choose a point in (0, 1) at random then the chance that the chosen number is less than 1/2 is 50%, while if all the numbers in (0, 1) are squared and we choose from among these squares uniformly, the chance will be only 25%. Of course, the first answer, i.e., 50% is more reasonable. However, in other problems it might be more difficult to choose between reasonable and unreasonable. We have already mentioned (in the last remark to the first paradox) that such a choice is not always possible on the basis of pure logic excluding experience. Exactly this is the essence of the following paradox published in the book "Calcul des probabilités" (1889) by *Joseph Louis Bertrand.*

b) The paradox

Choose a random chord of a given circle and calculate the probability that this chord is longer than the side of the equilateral triangle inscribed in the circle. The paradox claims that this probability is not determined uniquely, i.e., different methods lead to different results.

First method:
Choose a point at random, uniformly in the given circle. This random point determines a unique chord whose midpoint is the randomly chosen point. This chord is longer than the side of our equilateral triangle if and only if the point is interior to the inscribed circle of the triangle. The radius of this circle is half of the original one, that is, its area is 1/4 of the other. Consequently the probability that the randomly chosen point is in the inside of the inscribed circle is 1/4. So this method gives the answer 1/4.

Second method:
Due to symmetry, one end of the chord can be any fixed point on the circumference of the circle. So fix it to a vertex of the inscribed triangle. Choose the other end at random with uniform distribution. The vertices of the triangle divide the circumference of the circle into 3 equal arcs

and the random chord is longer than the side of the equilateral triangle if the random chord intersects the triangle. So the probability in question is now 1/3.

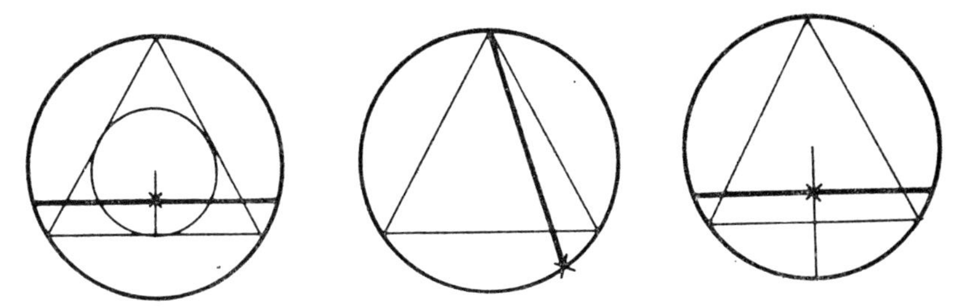

Figure 5. Three ways for choosing the random chord.

Third method:
Choose a point at random, uniformly on a radius of the circle and take the chord which is perpendicular to the radius at this point. Then the random chord is longer than the side of the inscribed equilateral triangle if the random point belongs to the half of the radius which is closer to the centre. Due to symmetry, it does not matter which radius was originally chosen, therefore the probability is 1/2.

c) The explanation of the paradox

The different results were considered a paradox since it was believed that "the uniform random choice" uniquely determines the probability in question. The paradox points out that there can be different uniform choices, all of which are "natural" in a sense. Each of the 3 methods above uses a uniform distribution (in the circle, on the circumference of the circle, and on a radius of the circle). In Poincaré's opinion (Calcul des probabilités, Paris, 1912) if we do not have any preliminary information then we should accept the third method (were the result is 1/2) because this is the method which assures that if two sets of chords are geometrically congruent then there is the same probability that a randomly chosen chord belongs to one set or the other. The study of this kind of invariance led to a very interesting branch of mathematics called integral geometry. (This term was coined by *Wilhelm Blaschke* in 1934.)

The following invariance requirements also lead to probability 1/2 (see Janes, E. T., "The Well-posed problem", *Foundations of Physics*, **3**, 477—493, (1973)). Let the circle have radius R. The position of the chord is determined by giving the polar coordinates (r, ϑ) of its center. We seek to answer a more detailed question than Bertrand's: What probability density $f(r, \vartheta)\, dA = f(r, \vartheta) r dr \vartheta$ should we assign over the interior area of the circle? Since the distribution of chord length depends only on the radial distribution, $f(r, \vartheta) = f(r)$. Thus the problem is reduced, determining a function $f(r)$, normalized according to

$$\int_0^{2\pi} \int_0^R f(r)\, r dr d\vartheta = 1, \quad \text{i.e.} \quad 2\pi \int_0^R f(r)\, r dr = 1.$$

The *scale invariance* (i.e., the invariance under the change of scale) leads to the equation

$$a^2 f(ar) = 2\pi f(r) \int_0^{aR} f(u) u du, \quad 0 < a \leqq 1, \quad 0 \leqq r \leqq R.$$

Differentiating with respect to a, setting $a=1$, and solving the resulting differential equation, we find that the most general solution (satisfying the above mentioned normalizing condition) is

$$f(r) = \frac{q r^{q-2}}{2\pi R^q}$$

where q is a constant in the range not further determined by scale invariance. Finally, if we translate the circle by a distance b the transformation $(r, \vartheta) \to (r', \vartheta')$ is given by $r' = |r - b \cos \vartheta|$ and

$$\vartheta' = \begin{cases} \vartheta & \text{if} \quad r > b \cdot \cos \vartheta \\ \vartheta + \pi & \text{if} \quad r < b \cdot \cos \vartheta. \end{cases}$$

The *translational invariance* gives $q=1$.

Thus we get

$$f(r, \vartheta) = \frac{1}{2\pi R r}; \quad 0 < r \leqq R, \quad 0 \leqq \vartheta \leqq 2\pi$$

corresponding to the third method. Since a chord whose midpoint is at (r, ϑ) has a length $L=2(R^2-r^2)^{1/2}$, the probability density function of

$$X = \frac{L}{2R} \quad \text{is} \quad \frac{x}{(1-x^2)^{1/2}}, \quad 0 \leqq x < 1$$

in agreement with Borel's conjecture (Élements de la théorie des probabilités, Paris, 1909).

d) Remarks

(*i*) In discussing Bertrand's paradox, we have dealt with three methods of choosing a random chord but there exist many other natural methods as well. E.g., if we pick a point in the given circle at random and draw a chord of any direction through the chosen point (the direction is uniformly distributed in the whole angular domain and independent of the choice of the point) then the probability in question is

$$\frac{1}{3} + \frac{\sqrt{3}}{2\pi} = 0.609\ldots.$$

It is not surprising that the result is more than 1/2 because this kind of selection prefers the longer chords. The probability is even greater (0.7449) if the random chord connects two random points of the circle. Another, less natural, method of choice is the following. Draw a concentric circle (with radius r) to the given circle (of radius R) and choose a random point (uniformly distributed) in the circle of radius r. Draw a line through this chosen point with a direction uniformly distributed in the whole angular domain and independent of the chosen point. Now the question is the following. If the line intersects the circle of radius R what is the probability that the chord which was cut out of the circle of radius R is longer than the frequently mentioned side of the inscribed equilateral triangle? The answer comes easily. If r is increased gradually from $r=\frac{R}{2}$ to $r=\infty$, the probability in question decreases from 1 to 1/2.

(*ii*) The integral geometry developed from geometric probability has an increasing importance in many fields, e.g., in stereology, in the reconstruction of 3-dimensional forms from their 2-dimensional sections or projections. Stereology is usefully employed in minerology, metallurgy, and biology (especially in tomography in the 3-dimensional reconstruction of tumours).

e) References

Ainley, E. S. "A probable paradox", *Math. Gaz.*, **66**, 300—301, (1982).

Kendall, M. G., Moran, P. A. P., *Geometric Probability*, Griffin, London, 1963.

Matheron, G., *Random Sets and Integral Geometry*, Wiley, New York, 1975.

Santaló, L. A., *Integral Geometry and Geometric Probability*, Addison-Wesley, Reading, 1976.

de Temple, D. W., Robertson, J. M., "Constructing Buffon curves from their distributions", *The American Math. Monthly*, **87**, 779—784, (1980).

Zalcman, L., "Offbeat integral geometry", *The American Math. Monthly*, **87**, 161—174, (1980).

Schuster, E. F., "Buffon's needle experiment", *The American Math. Monthly*, **81**, 26—29, (1974).

12. A PARADOX OF GAME THEORY. THE GLADIATOR PARADOX

a) The history of the paradox

Though gambling has flourished in various forms from the time of Paleolithic men and the mathematical study of various games goes back to the Renaissance, it was only in the 20th century that the general theory of games (and its connection with other sciences like economics) evolved. In 1921 a mathematical theory of game strategies was first attempted by *Emile Borel*, but it was *John von Neumann*, the father of game theory, who proved the *minimax principle*, the fundamental theorem in game theory in 1928. (Earlier even Borel had doubted its validity.)

The following paradox helps to understand the essence of the minimax theorem.

b) The paradox

Two children, R and Q, play the following wellknown children's game. They both put up one or two of their fingers at the same time; if the number of fingers they raised altogether is even, Q pays R, and if it is odd, R pays Q the same amount as the number of fingers raised altogether. The table (payoff matrix) below indicates the sum of money Q has to give to R. Though this game is generally considered to be fair [perhaps because the numbers shown in the table add up 0: $2+(-3)+(-3)+4=$ $=0$], it is not fair at all: it is definitely favourable for Q.

R \ Q	1 finger	2 fingers
1 finger	$ 2	$ -3
2 fingers	$ -3	$ 4

Figure 6.

c) The explanation of the paradox

Obviously, if one of the players always puts up one finger or always two, then, having observed this, his opponent can play so that he always wins. Therefore only a "mixed strategy" can be advantageous, that is, in each trial the player has to choose at random, but with fixed probabilities, from the two possibilities (one or two of his fingers). Let us suppose that we have already determined the optimal strategies of both players, i.e., we know that the best strategy for R is to put up one finger with probability p_1 and put up two fingers with probability p_2 (clearly $p_1+p_2=$ $=1$) and similarly for Q the most favourable is to lift up one finger with probability q_1 and two fingers with probability q_2 ($q_1+q_2=1$). Since the two players decide independently of each other, the average amount of money Q pays R (if both players have chosen the optimal strategy) is

$$V = 2p_1q_1-3p_1q_2-3p_2q_1+4p_2q_2. \qquad (*)$$

The game would be fair if $V=0$. We shall show, however, that $p_1=q_1=$ $=\frac{7}{12}$, $p_2=q_2=\frac{5}{12}$ and then $V=-\frac{1}{12}$, which means that Q wins, on the average, $\frac{1}{12}$ dollars in each trial even if R follows his optimal strategy.

Substitute $q_1=1, q_2=0$ in (*). Then $V=Q_1=2p_1-3p_2$. Similarly if $q_2=1$ and $q_1=0$, then $V=Q_2=-3p_1+4p_2$. Using this notation $V=$ $=q_1Q_1+q_2Q_2$. As V is the average loss of Q, if he follows his optimal strategy, $Q_1 \geqq V$ and $Q_2 \geqq V$, hence $V=q_1Q_1+q_2Q_2 \geqq q_1V+q_2V=$ $=(q_1+q_2)V=V$.

Since neither q_1 nor q_2 can be equal to zero, it follows from the above relation that $V=Q_1=Q_2$, i.e., $2p_1-3p_2=-3p_1+4p_2$, so (using p_1+ $p_2=1$) $p_1=\frac{7}{12}$, $p_2=\frac{5}{12}$ and $V=-\frac{1}{12}$. Similarly $2q_1-3q_2=-3q_1+$ $+4q_2$ $(q_1+q_2=1)$, and consequently $q_1=\frac{7}{12}$, $q_2=\frac{5}{12}$.

Thus we have proved that the game is certainly not fair, and we have also obtained the optimal strategy. For both players it is advantageous to raise one finger with probability $\frac{7}{12}$.

Substituting $1-p_1$ for p_2 and $1-q_1$ for q_2 in the formula (*): $V=$ $=12p_1q_1-7p_1-7q_1+4$. For $p_1=\frac{7}{12}$, $V=-\frac{1}{12}$ regardless of the value of q_1; similarly for $q_1=\frac{7}{12}$, $V=-\frac{1}{12}$ regardless of the value of p_1. Accordingly, it makes no difference to a player how he plays if he knows that his opponent has chosen his optimal strategy.

d) Remarks

(*i*) The principal aim of Neumann's research in game theory was to find the optimal strategy of a game in which m players take part. We assume for simplicity that $m=2$ (i.e., only two players play against each other) and that the game has the zero-sum property (i.e., the loss of the first player is equal to the gain of the second player). Let S_1 and S_2

denote the sets of the first and the second player's pure strategies, respectively (a pure strategy is a rule which determines the first player's first step and the replies to all the possible steps of the opponent). Let $L(s_1, s_2)$ be a bivariate function which gives the loss of the second player if he follows the pure strategy $s_2 \in S_2$ and the first player follows $s_1 \in S_1$ (the table on page 49 shows a function of this type). For a cautious player the best strategy is the one which minimizes his maximum loss (which occurs at optimal defence). The first player can manage to win a gain

$$V_1 = \max_{s_1} (\min_{s_2} L(s_1, s_2))$$

anyway, and the second player

$$V_2 = \min_{s_2} (\max_{s_1} L(s_1, s_2)).$$

(Naturally V_1 or V_2 may also take negative values indicating actual loss.) In the case where $V_1 = V_2$ and the sets of possible strategies are finite, it is useful for both players to choose the strategies s_1^*, s_2^* for which $V_1 = V_2 = L(s_1^*, s_2^*)$. A strategy-pair such as (s_1^*, s_2^*) is the saddle-point of the game, but it does not always exist. Neumann, however, had the brilliant idea of extending the set of possible strategies and introduced "mixed-strategies" which choose randomly from pure strategies. Thus a mixed strategy is a probability distribution on the set of pure strategies. (In the children's game example the mixed strategies of the two players were p_1, p_2 and q_1, q_2, respectively.) Mixed strategies eliminate the possibility of a player "seeing through his opponent", but it introduces chance, even in games where the rules themselves do not depend on chance. Naturally, if we want to find the optimal mixed strategies then we have to define the loss function on the set of pairs (π_1, π_2) of mixed strategies. Let $L(\pi_1, \pi_2)$ be the average loss that the second player pays the first one if they choose the mixed strategies $\pi_1 \in P_1$ and $\pi_2 \in P_2$. Neumann's minimax theorem (the fundamental theorem of game theory) states that if S_1 and S_2 are finite sets, then

$$\max_{\pi_1 \in P_1} \min_{\pi_2 \in P_2} L(\pi_1, \pi_2) = \min_{\pi_2 \in P_2} \max_{\pi_1 \in P_1} L(\pi_1, \pi_2),$$

i.e., there always exists a saddle-point in mixed strategies, thus optimal mixed strategies exist for both players.

The general model of game theory can be used to examine conflicts appearing in other fields of life, too. E.g., from a mathematical point of view commercial competition can be considered a "game" in which both players want to find their optimal strategies. Since it is less and less likely that rivals could swindle each other permanently, compromises (corresponding to saddle points) are becoming more and more important in many fields. Game theory brought a new aspect into mathematical statistics, too, mainly due to *Abraham Wald.* The following remark shows a few applications of game theory in statistics.

(*ii*) A typical problem in statistics is the estimation of the unkown parameter $\vartheta\in\Theta$ of a probability distribution F_ϑ, on the basis of the (usually independent) F_ϑ distributed observations $X_1, X_2, \ldots, X_n$, i.e., the sample (Θ is an arbitrary set usually consisting of numbers or vectors). Consider a bivariate function $L(\vartheta, c)$ the values of which mean our loss in the case where we estimate the unknown parameter ϑ with the value c. It is quite natural to assume that the greater the deviation $|\vartheta-c|$ is, the greater the loss becomes, thus $L(\vartheta, c)$ is typically a monotone increasing function of $|\vartheta-c|$, for example, $L(\vartheta, c)=|\vartheta-c|^\delta$, where $\delta>0$. An estimator $\hat{\vartheta}=f(X_1, X_2, \ldots, X_n)$ is good if the average loss is small, i.e., if the risk function $R(\vartheta, \hat{\vartheta})=E(L(\vartheta, \hat{\vartheta}))$ is small. Comparing two estimators, however, the value of the risk function of the first estimator at particular values of ϑ can be smaller than that of the second estimator, while at other values of ϑ the situation is quite the opposite. A comparatively wide range of estimators have a risk function which can be decreased at some values of ϑ only if at other values of ϑ it increases. These kinds of estimators are called *admissible* estimators, that is, an estimator $\hat{\vartheta}_0$ is admissible if the inequality $R(\vartheta, \hat{\vartheta})\leqq R(\vartheta, \hat{\vartheta}_0)$ holds for all $\vartheta\in\Theta$ if and only if $R(\vartheta, \hat{\vartheta})=R(\vartheta, \hat{\vartheta}_0)$ for all $\vartheta\in\Theta$. Only admissible estimators are worth using because for a non-admissible estimator we can always find another estimator the risk function of which is nowhere larger and definitely smaller at certain points than the risk function of the non-admissible estimator. If we want to find an admissible estimator which minimizes the average loss at the "worst" actual parameter value (where the risk function takes its maximum), then we obtain the minimax estimator. This cautious estimator is defined as follows: an estimator ϑ^* is called minimax if

$$\sup_{\vartheta\in\Theta} R(\vartheta, \vartheta^*) = \inf_{\hat{\vartheta}} \sup_{\vartheta\in\Theta} R(\vartheta, \hat{\vartheta}),$$

where $\hat{\vartheta}$ runs through all the possible estimators. The "minimax-aspect" of mathematical statistics considers estimation a game in which Nature "chooses" a parameter ϑ and we choose an estimator $\hat{\vartheta}$. The aim of the game is to make the average loss as small as possible. The average loss can be made less by allowing mixed strategies, when Nature chooses the parameter ϑ randomly from Θ with distribution τ and we also choose the estimator randomly with distribution α from the set of all possible estimators. In this case the risk function is $r(\tau, \alpha)=E\big(R(T, A)\big)$, where T is a τ distributed random variable on the parameter set Θ and A has the distribution α on the set of all possible strategies. The minimax theorem remains valid for risk functions of this kind under quite general conditions:

$$\sup_{\tau} \inf_{\alpha} r(\tau, \alpha) = \inf_{\alpha} \sup_{\tau} r(\tau, \alpha).$$

Since the distribution τ is unknown, it is useful to choose a mixed minimax strategy α^* as an estimator for which the equation

$$\sup_{\tau} r(\tau, \alpha^*) = \inf_{\alpha} \sup_{\tau} r(\tau, \alpha)$$

holds.

(*iii*) The following gladiator paradox comes from *K. S. Kaminsky, E. M. Luks* and *P. I. Nelson.* In a contest, called the gladiator game, suppose that two teams of gladiators are to do battle in the arena. In successive rounds a gladiator is selected from team $A=(A_1, A_2, \dots, A_n)$ to meet a gladiator selected from team $B=(B_1, B_2, \dots, B_n)$. The victor returns to his team with undiminished vigour to fight again, if needed. The looser, presumably disabled, is removed from the tournament. Individual matches are assumed to have a stochastic component and represent mutually independent trials where we let $0<P(A_i, B_j)<1$, denote the probability that gladiator A_i defeats gladiator B_j. The matches continue until one team is eliminated. We investigate the existence of strategies S which are optimal in the sense of maximizing $P_S(A, B)$, the probability that team A defeats team B when strategy S is used. A strategy here is a rule which decides the order in which gladiators from both teams enter the arena. (Only the current composition of the teams can be used in formulating the strategy at each stage of the game.) In a special gladiator game let $a_1, a_2, \ldots, a_n, b_1, b_2, \ldots, b_n$ be positive strengths assigned to $A_1, A_2, \ldots, A_n, B_1, B_2, \ldots, B_n$, respectively, such that for all contests A_i vs B_j, $P(A_i, B_j)=a_i/(a_i+b_j)$. Then the probability $P_S(A, B)$ is the same for all S! This is the gladiator paradox. Another paradox of this game is the following. Say that A dominates B if $P(A, B)>1/2$. Now if A dominates B and B dominates C then A does not necessarily dominate C. There are examples showing that $m=\min\{(P(A, B), P(B, C), P(C, A))\}>1/2$, though the upper limit of m is an intriguing open question. (A related paradox is I/13f.)

(*iv*) A game theoretical paradox is the famous "prisoner's dilemma". Here we only refer to the paper by Brams, Straffin, and Hofstadter.

e) References

Brams, S. J., Straffin, Jr. S. J., "Prisoners dilemma and professional sport drafts", *The Amer. Math. Monthly*, **86**, 80—88, (1979).

Hofstadter, D. R., "Computer tournaments of the Prisoner's dilemma suggest how cooperation evolves", *Sci. Amer.*, 16—26, (May 1983).

Joó, I., "A simple proof for Neumann's minimax theorem", *Acta Sci. Math.* (Szeged), **42**, 91—94, (1980) and *Acta Math. Hung.* **44**, 363—365, (1984).

Neumann, J., Morgenstern, O., *Theory of Games and Economic Behavior*, Princeton Univ. Press, Princeton, 1944.

Savage, L. J., *The Foundations of Statistics*, Wiley, New York, 1954.

Wald, A., *Statistical Decision Functions*, Wiley, New York, 1954.

Williams, J. D., *The Complete Strategyst Being a Primer on the Theory of Games of Strategy*, McGraw-Hill Book Company, New York.

13. QUICKIES

a) The paradox of "almost sure" events

Consider two random events with probabilities of 99% and 99.99%, respectively. One could say that the two probabilities are nearly the same, both events are almost sure to occur. Nevertheless the difference may become significant in certain cases. Consider, for instance, independent events which may occur on any day of the year with probability $p=99\%$; then the probability that it will occur every day of the year is less than $P=3\%$, while if $p=99.99\%$ then $P=97\%$.

b) The paradox of probability and relative frequency

The following story, from *George Polya,* shows how not to interpret the frequency concept of probability. D. Tel (doctor of teleopathy) shook his head as he finished examining his patient. "You have a very serious disease," he said, "of ten people who have got this disease only one survives". As the patient was sufficiently scared by this information, D. Tel began to console him: "But you are very lucky sir, because you came to me. I have already had nine patients who all died of it, so you will survive."

(Ref.: Polya, G., *Patterns of Plausible Inference*, Vol. II, Princeton Univ. Press, 1954, p. 101.)

c) Coin paradoxes

(*i*) We toss a fair coin until we score two heads (HH) or a head and a tail (HT) in succession. Obviously the probability that (HH) will occur sooner than (HT) is equal to the probability that (HT) will occur sooner since after tossing a H the coin still falls H or T with equal probability. In spite of this fact more tosses are necessary, on average, for (HH) than for (HT) to turn up. (HH) occurs in 6 tosses and (HT) in 4 tosses, on average. [Let M_H denote the expected value of the number of tosses we need to score (HH) assuming that H has occurred in the first toss, and let M_T denote the expected value of number of tosses necessary to score (HH) assuming that T has occurred in the first toss. Then $M_H=1+$ $+(1+M_T)/2$ and $M_T=1+(M_H+M_T)/2$, and it follows that $(M_H+M_T)/$ $/2=6$, i.e., the average number of tosses necessary to score (HH) is indeed 6.] The contrast is sharper if we compare the sequences $(HTHT)$ and $(THTT)$. The probability that $(HTHT)$ will appear sooner than $(THTT)$ is $\frac{9}{14}>\frac{1}{2}$, but the average number of tosses we need to score

$(HTHT)$ is still greater than that of the tosses necessary to get $(THTT)$. (The former is 20 whereas the latter is only 18.) Thus even if the probability that the event A will occur sooner than B is larger than the probability that B will occur sooner, we may still have to wait more, on average, for A than for B.

Incidently, it can be proved that among the $H-T$ sequences of length n the pure sequence has the longest expected waiting time (i.e., which consists either of n H's of n T's). In this case the expected number of tosses is $2^{n+1}-2$. The smallest possible (average) number of tosses is 2^n, which occurs when we want a sequence consisting of $n-1$ $H's$ in succession followed by one T, or $n-1$ $T's$ followed by one H. (Thus we have to wait almost twice as long for the head run of length n than for the sequence of $n-1$ $H's$ and one T, although the probability that the former will appear sooner equals the probability that the latter will appear sooner.)

Determining the length of the time-interval, we have to wait for a given $H-T$ sequence of length n to appear, usually requires cumbersome calculation (solving multivariate linear system of equations) for large n. Calculations can be considerable simplified by using the "magic" Conway algorithm, which is discussed in the article by *Li* quoted below. We shall give it now in a more general form not only for fair $H-T$ sequences. Let $X, X_1, X_2, \ldots,$ be independent, identically distributed random variables assuming only a finite or countably infinite number of values with positive probabilities. Denote the set of these values by V and let $A=(a_1, a_2, \ldots, a_m)$ and $B=(b_1, b_2, \ldots, b_n)$ be two (finite) sequences whose elements are from V. Introduce the following notation:

$$d_{ij} = \begin{cases} \dfrac{1}{P(X=b_j)}, & \text{if } 1 \leqq i \leqq m, \quad 1 \leqq j \leqq n \text{ and } a_i = b_j \\ 0 & \text{otherwise} \end{cases}$$

and $A \cdot B = d_{11} d_{22} \ldots d_{mm} + d_{21} d_{32} \ldots d_{m,m-1} + \ldots + d_{m1}$. Let T_A and T_B denote, respectively, the number of X-variables until the first occurrence of the seqeunce A and B in $X_1, X_2, \ldots$. Then the expected value of T_A is $A \cdot A$, while the probability that T_A is smaller than T_B (assuming that neither A nor B contains the other as a connected block) is

$$\frac{B \cdot B - B \cdot A}{A \cdot A + B \cdot B - A \cdot B - B \cdot A}.$$

For example, in a fair $H-T$ sequence $d_{ij}=0$ or $d_{ij}=2$; if $A=(HTHT)$ and $B=(THTT)$, then $A \cdot A=20, B \cdot B=18, A \cdot B=10$ and $B \cdot A=0$, so the probability that A occurs sooner than B is $\frac{18}{28}=\frac{9}{14}$ as we have mentioned above.

Finally we state a more sophisticated (still unpublished) theorem. If we wait until all the possible 2^n $H-T$ sequences of length n occur (tossing a fair coin) and τ_n denotes the (random) waiting time then

$$\lim_{n \to \infty} P(\tau_n/2^n - \log 2^n < x) = e^{-e^{-x}}.$$

(*ii*) A fair coin has to be tossed, on average, at least 8 times if we want a given sequence of length 3 (e.g., HHT). The number of necessary tosses is the smallest if we want to score any of the following sequences: (HHT), (THH), (TTH), (HTT). (In each case the average number of necessary throws is 8, whereas in any other case it is more than 8.) Compare these sequences in the following way:

α) the probability that (HHT) will occur sooner than (THH) is $\frac{1}{4}$,

β) (THH) will appear sooner than (TTH) with probability $\frac{1}{3}$,

γ) (TTH) will occur sooner than (HTT) with probability $\frac{1}{4}$, and finally

δ) the probability that (HTT) will occur sooner than (HHT) is $\frac{1}{3}$.

Thus, having started from the sequence (HHT), we have reached (HHT) again, though in each step the comparative probabilities were strictly less than 1/2.

(Ref.: *Li, Shou-Yen R.* "A martingale approach to the study of occurrence of sequence patterns in repeated experiments", *Annals of Probability*, 8, 1171-1176, (1980).)

d) The paradox of conditional probability

Events A, B and C exist such that

α) the conditional probability of A given B is smaller than the conditional probability of A given that B has not occurred;

β) the conditional probability of A given that both B and C have occurred is larger than the conditional probability of A given C has occurred but B has not occurred;

γ) the conditional probability of A given B and the complement of C is larger than that of A given that neither B nor C have occurred.

Using symbols the three statements can be written as follows:

$$P(A|B) < P(A|\bar{B}),\ P(A|BC) > P(A|\bar{B}C)$$

and

$$P(A|B\bar{C}) > P(A|\bar{B}\bar{C}).$$

This seems to be paradoxical because one might think that $P(A|B)$ is the average of $P(A|BC)$ and $P(A|B\bar{C})$ and, similarly, that $P(A|\bar{B})$ is the average of $P(A|\bar{B}C)$ and $P(A|\bar{B}\bar{C})$, and the average of two smaller values must be smaller than the average of two larger values. The explanation of this misconclusion is that $P(A|B)$ and $P(A|\bar{B})$ are the weighted average of the above mentioned probabilities but the respective weights are different in the two cases:

$$P(A|B) = P(C|B)\,P(A|BC)+P(\bar{C}\,|B)\,P(A|B\bar{C}),$$

whereas

$$P(A|\bar{B}) = P(C\,|\bar{B})\,P(A|\bar{B}C)+P(\bar{C}\,|\bar{B})\,B(A|\bar{B}\bar{C}).$$

Nevertheless if the events B and C are independent then $P(C|B)= =P(C|\bar{B})$ and $P(\bar{C}|B)=P(\bar{C}|\bar{B})$, so in this case the paradoxical phenomenon cannot occur.

(Ref.: *Blyth. C. R.* "On Simpson's paradox and the sure thing principle." *J. Amer Statist. Assoc.* 67. 364-366. (1972).)

e) The paradox of random waiting times

Two random events occur after a (random) time X and Y. Paradoxically, it may happen, that $X>Y$ with a probability of at least 99%, but X is stochastically smaller than Y, i.e., the probability of $X<t$ is larger

than the probability of $Y<t$ for any fixed time t (or, in other words, the distribution function of X is everywhere larger than that of Y). E.g., if Y is uniformly distributed in the interval $[0, 1]$, $X=Y+(1-Y)/1000$ with probability 99% and $X=Y/1000$ with probability 1%. [This paradoxical situation cannot occur if X and Y are independent: let F and G denote their distribution functions; for simplicity assume that G is continuous and its inverse function G^{-1} exists. Then the distribution function of the random variable $Z=G^{-1}(F(X))$ is also G. Since $F>G$, $Z>X$, hence

$$P(X>Y) \leqq P(Z>Y) = \frac{1}{2}$$

as Z and Y are identically distributed independent random variables, i.e., $P(X>Y)$ must be much smaller than 99%, in fact, not more than 50%].

The following paradox is similar to the preceding one. Let X and Y be two independent random variables such that X is stochastically smaller than Y. Then one might think that $\max(X, X+Y)$ is also stochastically smaller than $\max(Y, X+Y)$, but that is not true, for example, in the case where X and Y both may assume only the values -1, 0 and 1 with probabilities $\frac{1}{4}, \frac{1}{4}, \frac{1}{2}$ and $\frac{1}{4}, \frac{1}{2}, \frac{1}{4}$, respectively.

(Ref.: Blyth, C. R., "Some probability paradoxes in choice from among random alternatives" (with comments by D. V. Lindley, I. J. Good, R. L. Winkler, and J. W. Pratt), *J. Amer. Stat. Assoc.*, **67**, 366—388, (1972). See also *SIAM Rev.*, April 1970.)

f) The paradox of transitivity

Two players, A and B, are playing the following game. In the first step A numbers 3 dice to his taste, writing one of the numbers 1, 2, 3, ..., 18 on each face of the 3 dice (he must use each number only once.)

In the second step B scrutinizes the 3 dice (numbered by A) and chooses one of them.

In the third step A chooses one of the remaining 2 dice. In the last step both A and B throw their dice and the player who scores the larger number wins.

One might think that this game is more favourable to B, because no matter how A numbers the 3 dice B can always choose the best one (or one of the best ones), thus B has the chance of at least 50% of winning. But, paradoxically, just the opposite is true: A can number the dice so that he wins with probability $\frac{21}{36}$ (which is more than 50%), no matter which dice B chooses. This is because of the "round defeat" numbering system where each dice defeats exactly one of the other two, which means there is no "best" among the dice. Let I, II and III denote the three dice and suppose that A numbered the dice in the following way. He wrote the numbers

	18, 10, 9, 8, 7, 5	on the faces of dice I,
	17, 16, 15, 4, 3, 2	on the faces of dice II,
and	14, 13, 12, 11, 6, 1	on the faces of dice III.

It can be easily calculated that we get a larger number with dice I than with dice II with probability $\frac{21}{36}$, similarly, a larger number is scored with dice II than with dice III with probability $\frac{21}{36}$, and the probability that a larger number appears on dice III than on dice I is also $\frac{21}{36}$; therefore the "round defeat" probability is $\frac{21}{36}$. So if A numbers the dice this way, he is in a more favourable position than B. (If B chooses the dice I, II or III and, accordingly, A chooses the dice III, I or II, A has more chance of winning.) It can also be proved that the probability of "round defeat" cannot exceed $\frac{21}{36}$. The point of this paradox is that random quantities may not be ordered according to which one is larger than the other with a probability of more than 50%, because the transitivity does not hold. If the same number may be written on more than one face of the dice, e.g.,

1, 4, 4, 4, 4, 4 on dice I,

2, 2, 2, 5, 5, 5 on dice II,

and 3, 3, 3, 3, 3, 6 on dice III,

then the probability of "round defeat" is also $\frac{21}{36}$. Formulate the paradox more generally. Let $X_1, X_2, \ldots, X_n$ be arbitrary numbers depending on chance (i.e., random variables). Denote the probabilities of the events $X_1 < X_2$; $X_2 < X_3$; ...; $X_n < X_1$, by $p_1, p_2, \ldots, p_n$, respectively, then $\min(p_1, p_2, \ldots, p_n)$ is the probability of "round defeat". Let k_n denote this probability. The larger k_n is the sharper the paradox becomes; it can easily be shown than k_n can never exceed $\frac{n-1}{n}$ and this is the least upper bound. The calculation of the least upper bound of k_n is more difficult if the random variables $X_1, X_2, \ldots, X_n$ are supposed to be independent as the outcomes of rolling dice. Let f_n denote the least upper bound in this case. *Usiskin* calculated (*Annals of Statist.*, **35**, 857—862, 1964) that $f_2 = \frac{1}{2}$, $f_3 = \frac{\sqrt{5}-1}{2}$ (the ratio of golden section), $f_4 = \frac{2}{3}$, etc. The sequence $\{f_n\}$ increases monotonically and converges to $\frac{3}{4}$. One can also show that the speed of convergence is of order n^{-2}.

g) The paradox of measurement for regularity of dice

In dice throwing the same face will appear twice in succession in 7 throws, on average, and three times in succession in 43 throws (see the end of this paradox). If the dice were biased (i.e., different faces appeared with different probabilities), the average number of necessary throws to get the same face twice or three times in succession would be smaller. We shall call dice I more regural than dice II if dice I has to be thrown more times on the average to get the same face twice (or three times) in succession than dice II. Paradoxically, more throws may be necessary on average with dice I than with dice II to score the same face twice in succession, but to score the same face three times in succession, dice II has to be thrown more times.

The following simple example comes from *T. F. Móri*. Suppose that each face of dice I can be scored with probabilities 0.03; 0.03; 0.19; 0.19; 0.28; 0.28 and let the corresponding probabilities with dice II be 0.04; 0.04; 0.17; 0.17; 0.29; 0.29. Then dice I has to be thrown 5.41 times and dice II 5.47 times on the average to score a face twice in succession, whereas if we want to score a face three times in succession then we have to throw dice I 22.54 times and dice II 22.35 times on an average. This paradox shows that it is not expedient to define the "regularity" of a dice as we did. (In general it can be proved that if a particular face of a dice appears with positive probability p, then the average number of necessary throws to score this face k-times in succession is $m_p = p^{-1} + p^{-2} + \ldots + p^{-k}$. Consider a dice whose faces appear with probabilities $p_1, p_2, \ldots$ and let M_k denote the average number of throws we need to score the same face k-times in succession. Then $M_k^{-1} = m_{p_1}^{-1} + m_{p_2}^{-1} + \ldots$. If we put $p_1 = p_2 = \ldots = \frac{1}{6}$, then $M_2 = 7$ and $M_3 = 43$ as we have already mentioned.)

h) The birthday paradox

If not more than 365 people come together, it is possible that everybody has a different birthday, while with 366 persons it is certain (100%) that at least two of them were born on the same day of the year. (Let us ignore the existence of leap years here.) However, if we aim at 99% certainty, then, surprisingly, 55(!) people are enough to claim that there will be two among them having the same birthday, while for 68 people the probability that at least two of them have the same birthday is 99.9%. It is almost unbelievable that such a small difference between the probabilities 99% and 100% can lead to such a big difference between the number of people. This paradoxical phenomenon is one of the main reasons why probability theory is so wide-ranging in its application. (A similar phenomenon was mentioned in I. 10 Remark (*ii*).)

Denote by n the number of days in a year and by x $(<n)$ the number of people in a group. The probability that no two people in this group

have the same birthday is then

$$\frac{n(n-1)(n-2)\dots(n-x+1)}{n^x}.$$

Therefore if

$$\frac{n(n-1)(n-2)\dots(n-x+1)}{n^x}=1-p$$

then p is the probability that among x people there are some having the same birthday. The approximate solution of this equation (provided that $0<p<1$) is

$$x\approx\sqrt{2n\ln\frac{1}{1-p}}.$$

Hence the order of magnitude of x is $\sqrt{n}$ for any value of p in the open interval (0, 1), while for $p=1$ $x=n+1$. A generalization of the birthday problem is the following. Calculate the lower bound x so that in a group of x people there be at least k who have their birthday on the same day of the year with probability p. Here the result is

$$x\approx cn^{(k-1)/k}$$

where c is a constant depending only on p and k (more precisely $c=\left(k!\ln\frac{1}{1-p}\right)^{1/k}$).

i) The paradox of heads and tails

Suppose we are playing heads or tails with a fair coin and we toss it 100 times .Then, surprisingly, the probability of the event $A=\{$we toss exactly 50 heads$\}$ is bigger than the probability of $B=\{$we toss at least 60 heads$\}$.

[As we have mentioned in I.10, $P(A)\approx 8\%$, while according to the Moivre—Laplace theorem $P(B)=1-\Phi(2)\approx 3\%$. The chance of tossing at least 55 heads, however, increases to about 16%, which is the double of $P(A)$.]

j) The edge of the coin

Generally the occurrence of a coin falling on its edge is left out of consideration since this event almost never occurs. Calculate now the size of a coin that ensures the same $\left(\frac{1}{3}\right)$ probability for falling heads, tails, and edge. For simplicity, consider the coin a flat cylinder whose bases are the heads and tails and the nappe is the edge. If the coin is spun around an axis which goes through the centre of the coin and is parallel to its bases it is enough to consider a planar section of the coin which contains the centre of the coin and is perpendicular to both bases. This section is a rectangle. Draw a circle around this rectangle and choose a landing point at random on its circumference. It is reasonable to suppose that the coin falls on its edge with a probability which is equal to the chance that the radius connecting the centre and the random point on the circumference intersects the side of the rectangle which corresponds to the nappe of the coin. In this model the coin falls on its edge with probability of 1/3 if the rate of its thickness and diameter is equal to tg $30° \approx 0.577$. The problem is not reduced to a planar one if the coin may turn round freely, more precisely, if the random point is chosen on the surface of the sphere drawn around the coin and we suppose that the coin falls on its edge if the radius connecting the random point and the centre intersects the nappe of the coin. In this model, tossing an edge will have the same probability as tossing heads or tails if the rate of the thickness and the diameter is 0.354.... There are, of course, more realistic models, too. The most surprising of them is the one where the above rate is the least (i.e., where the coin is the flattest).

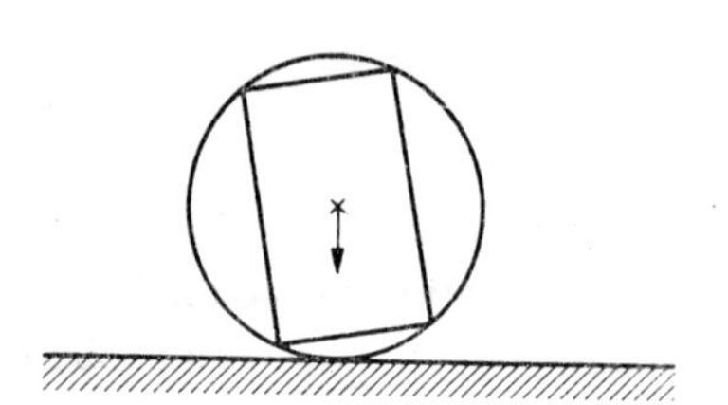

Figure 7. When does a coin fall on its edge?

k) Borel's paradox

Let a random point be chosen uniformly on the surface of a sphere (e.g., on the Earth, supposing its form is a sphere). The position of a point is generally given by its longitude and latitude. Given a latitude, the longitude is uniformly distributed, but given a longitude the distribution of the latitude is not uniform. (Its density function is proportional to the cosine of the latitude.) Consequently, the distribution of the random point is not the same if we suppose that it is on the equator or on the Greenwich meridian, though both the equator and the meridian are great circles on the globe and therefore their role seems to be symmetric.

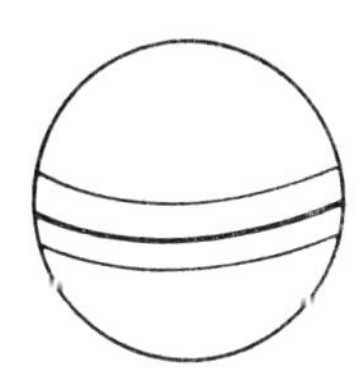

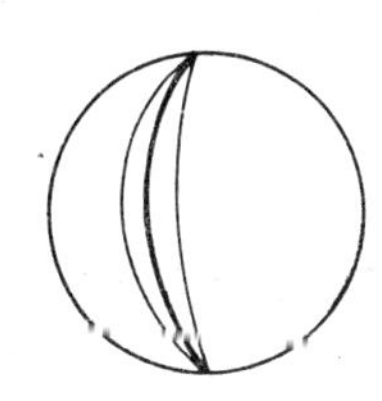

Figure 8. Though both the equator and a meridian are great circles on the globe, when calculating conditional probabilities, one should take into account the fact that while the equator is surrounded by spherical zones, a meridian is surrounded by biangles.

*

The next problem is a similar paradox. Let X and Y denote two independent normal distributions. (X, Y) can be considered a random point on the plane. Let R and φ be its polar coordinates. Supposing that $X=Y$ the distribution of $R^2=2X^2$ is the same as the distribution of the square of a standard normal random variable multiplied by 2. At the same time, under the condition $\varphi=\frac{\pi}{4}$ or $\varphi=\frac{5\pi}{4}$, the distribution of $R^2=X^2+Y^2$ is the same as that of the sum of the squares of two independent standard normal random variables (since R and φ are independent). Hence we get completely different distributions for R^2 in the case $X=Y$ and in the case $\varphi=\frac{\pi}{4}$ or $\varphi=\frac{5\pi}{4}$ which seems to be a paradox, because the two conditions mean just the same, only in the first case it is formulated by usual coordinates and in the second case by polar coordinates.

(Ref.: Billingsley, P., *Probability and Measure,* Wiley, New York—Chichester—Brisbane—Toronto, 1979).

l) A paradox of conditional distributions

Let X and Y be random variables and $f(x, y)$ a function of two variables such that for any fixed y the variable $f(X, y)$ is independent of Y. Is it true that in this case $f(X, Y)$ is also independent of Y? The following simple example shows that the answer is negative. Let $X=Y$ be uniformly distributed on the interval (0, 1). And let $f(x,y)=y$ if $x=y$ and $f(x,y)=0$ otherwise. Now $f(X, y)$ is indetically 0 (with probability 1), therefore it does not depend on Y, while $f(X, Y)=Y$ obviously depends on Y.

(Ref.: *Perlman, N. D. Wichura M. J.* A note on substitution in conditional distribution,, *Annals of Statist.* 3, 1175-1179. (1975).)

m) Winning a losing game

Suppose that in a game the number of trials (n) is always even. The first player A has a chance $p=0.45$ to get a point; for B, it is $p=0.55$. To win the game, one of them has to collect more than half of the points. If A has the privilege of fixing n then paradoxically $n=2$ is not the best choice. (This would be the best choice if p were very small, i.e., less than 1/3). If $p=0.45$ and $n=2$ then the probability that A wins is only $0.45^2=0.2025$, but if A has more trials he gets in a more favourable position. It is easy to prove that the optimal choice is $n=10$. This seems to contradict the general "principle" that the sooner we get out of a losing game the better. Suppose, e.g., that we need 20 dollars and have only 10. We want to win the missing sum by playing roulette. Since roulette is a losing game, it is adviceable to have as few tries as possible, i.e., we have to stake all of our money, e.g., on red. In this case the chance of winning is 18/38 (in American Roulette there are two zeros: 0 and 00). At the same time if we bet only one dollar in every trial we reach our aim with probability 0.11. For further details see Dubins, L. E., Savage, L. J., *How to Gamble if You Must,* New York, McGraw-Hill, 1965.

(Ref.: Mosteller, F., *Fifty Challenging Problems in Probability with Solutions*, Reading, Addison-Wesley, 1965.)

n) The paradox of insurance

A client, whose property is V, wants to insure a bV part of this property $(0<b<1)$ against a damage occurring with probability p each year. The annual premium is cV $(0<c<1)$. The insurance is effected by the company only if the expected value of its profit is positive, i.e., if c is greater than pb. Why do clients still insure if they know that it is profitable for the company and not for them. If the client insures and pays the money for n years, but the insurance company never has to pay then the client's initial property (V) will decrease to $V(1-c)^n$. And what happens if he does not insure? Let X_k denote the random variable which is equal to 1 if the client suffers a loss in the kth year and let $X_k=0$ otherwise. In this case his property in the $(k+1)$th year will be $V_{k+1}=$ $=V_k(1-bX_{k+1})$, therefore after n years

$$V_n = V\prod_{k=1}^{n}(1-bX_k) = Ve^{\sum_{i=1}^{n}\ln(1-bX_k)}.$$

Since the expected value of $\ln(1-bX_k)$ is $p\ln(1-b)$,

$$V_n \approx Ve^{np\ln(1-b)} = V(1-b)^{np}$$

with great probability. Thus the insurance is favourable for the client if $V(1-b)^{np}$ is less than $V(1-c)^n$, i.e., (using the expansion in power series) if c is less than

$$pb+\frac{p(1-p)}{2}b^2+\frac{p(1-p)(2-p)}{6}b^3+\ldots.$$

It means that the insurance is favourable for both the client and the company if c is greater than pb but less than the above sum. It is easy to see that the less b is (i.e., the smaller part of the client's property is insured) the less is the freedom in the choice of c, i.e., the possibility of compromise decreases. (In a sense, lottery is an insurance too; if somebody tips always the same numbers but stops playing after a while and his numbers were drawn afterwards, he would probably die of apoplexy. From this point of view, the prize of the tickets is really favourable. Football pools are another case for there are very few people who always give the same tips and therefore it is not obvious what he missed when he did not play.)

o) Absurdities, Lewis Carrol

We will finish the series of quickies with absurdities and fallacies. We will mention problems together with their nonsense solutions, but to find the mistake in the reasoning may cause some brain beating. Lewis Carrol, the famous writer, was very fond of absurdities both in mathematics and literature. (The Absurd Literature by Nicolae Balote considers Carrol the number one forerunner of modern absurdity.) In his last 10 years, Carrol was attracted by mathematical absurdities (see the collection of Curiosa Mathematica 1888 or the article of the Mind published in April 1895). In his Pillow Problems (1894) the following absurdity can be read.

There are two balls in a bag, they are either red or white. Let us guess their colour without looking into the bag. According to Carrol, the only correct answer is that one of them is red and the other is white. He gave the following reasoning. If there were 2 red (R) and 1 white (W) balls in the bag then the probability of drawing a red one would be 2/3. On the other hand, if there were 3 balls in the bag, and the probability of drawing a red one were 2/3, then there would be 2 R and 1 W balls in the bag.

Now put an R ball into the bag that originally contained only two balls. In this case there are four equally probable $\left(\frac{1}{4}\right)$ ball combinations: RRR, RWR, RRW and RWW. If the first combination is the actual one then the probability of drawing a R ball is 1, in the second and the third cases this probability is 2/3, and in the last one it is $\frac{1}{4}$. Therefore the probability of drawing a R ball is $1\cdot\frac{1}{4}+\frac{2}{3}\cdot\frac{1}{4}+\frac{2}{3}\cdot\frac{1}{4}+\frac{1}{3}\cdot\frac{1}{4}=\frac{2}{3}$. Thus there must be 2 R and 1 W balls in the bag, consequently there must have been 1 R and 1 W balls in the bag before we put a R ball into it. This result is obviously nonsense, so the reasoning must be false. But where is the mistake?

The following reasoning results in an absurdity as well. Two of three prisoners, denoted by A, B, C, will be executed. They know this, too, but cannot be sure who the lucky third will be. A says: "the probability that only I will not be executed is 1/3. If I ask the warder to tell me one of the 2 prisoners' names (different from mine) who will be executed then there remain only two possibilities. Either I am the other one to be executed or not, and therefore my chance for survival will increase to 1/2." However, it is also true that A knows even before the warder answers that one of his companions will certainly be executed and therefore the warder has not told him any essential information concerning his own execution. Why then has the probability of his execution changed?

(The answer is very simple: the probability has not changed at all, it has remained 1/3. The prisoner failed to take into account that the warder says, e.g., B with the probability of 1/2 if B and C are going to be executed, while if A and B are the victims, this probability is 1. Consequently, A's actual chance for escaping execution equals the ratio of the probability in the former case and that of the two cases together: $\frac{\frac{1}{6}}{\frac{1}{6}+\frac{1}{3}}=\frac{1}{3}$.)

Chapter 2

Paradoxes in mathematical statistics

"Statistics is the physics of numbers."

P. Diaconis

"Everything of importance has been said before by somebody who did not discover it."

A. N. Whitehead

"If one can tell ahead of time what one's research is going to be, the research problem cannot be very deep and may be said to be almost nonexistent."

A. Schild

Originally statistics was "state arithmetics". (The word statistics comes from the Latin *status*=state.) Since ancient times statistics have been applied to inform state leaders about the amount of taxes they can levy on their people and about the number of soldiers they can count on in war time. In China, a census was taken more than four thousand years ago. According to the Bible, Moses also counted all the men over 20 in his tribe. The result was 603,550. The fourth book of Moses (Book of Numbers) contains many other census data, but they seem to be exaggerated, as are the date of *Athenaios* giving the number of slaves in the Greek polices at the time of the Roman Empire. It is rather unlikely that there were 400,000 slaves in Athens and 460,000 in Corinth. We do not know how these census data have swollen, but it is a fact that, according to the census, the first city with more than a million inhabitants was Rome. England's first statistical document, the *Domesday Book* written in the 11th century, was also for purposes of taxation and the army. This is the reason why women have always been disregarded during a census right up until modern times. Statistics became a science only in the 17th century. Its pioneers were *John Graunt* (1620—1674) and *Sir William Petty* (1623—1687). Graunt's "Natural and Political Observations made upon the Bills of Mortality" (1662) was a demographic study. In 1669 Huygens published a life table based on Graunt's data. Petty's "Treatise on Taxes" (1662) and "Observation upon the Dublin Bills of Mortality" (1681) also used Graunt's results and ideas. In his "The Political Arithmetic" (published posthumously in 1689) Petty compared England, Holland, and France on their population, trade, and shipping. The term "political arithmetics" can be considered as the forerunner of the word "statistics". As capitalism advanced, not only state leaders but also capitalists became interested in statistical tables. More and more complicated mathematical means were used to process data, and their profit also increased, e.g., in the insurance business. Lloyd's, one of the outstanding insurance companies in the world, was founded in the 17th century, though at that time it was only a coffee-house in Tower Street in London. Good insurance is based on exact surveys and proper mathematical conclusions. Since the 17th century, mathematical statistics have gradually developed into an independent branch of mathematics. Its main purpose is to obtain as much correct and useful information as possible from the data, observations or measurements, in short from the *statistical sample.* (Measuring the amount of information apart from its concrete content developed into a new branch of mathematics only in the 20th century, and is called now information theory. It is very closely related to mathematical statistics.) Not to write satire, at least in *Juvenalis*' opinion, is hard, but not to find paradoxes in mathematical statistics is even harder. According to a joke, in 1901, 33% of the women students of Harvard University married their tutors. Actually, at that time only 3 girls studied at the university, and one of them did marry her professor. The statement is true, though misleading. Suppose that in a certain country 20% more boys than girls are admitted to the universities. If all the candidates are equally qualified for entry and the number of boy candidates is the same as the number of girl candidates then the obvious conclusion is that admission committees give preference to boys. However, since more girls than boys want to study at the more popular faculties, where the refusal rate is higher, the result may be that despite proportional admittance, there will be more boys studying at university than girls. *L. P. Ayres*' 1913 text analysis is similarly misleading or at least it is easy to misinterpret it. He states that the 50 most frequent words make up about 50% of a typical text, the 300 most frequent words make up 75%, while the 1000 most frequent words make up 90% of the text. In spite of this fact we should not conclude that if we know 50 or 100 words of a language we already understand half of it, for the knowledge of articles, though they are frequently used, can hardly help in understanding a text. No wonder many people believe there are three kinds of lies: white lies, damned lies, and statistics. We hope that the explanations of paradoxes in mathematical statistics will help us to see through statistical absurdities and to understandt the useful and essential conclusions of statistics as well as to find the most fundamental information.

1. BAYES' PARADOX

a) The history of the paradox

A student of de Moivre, *Thomas Bayes,* was one of the most outstanding pioneers of mathematical statistics. His theorem discovered about 1750 but published only after his death was the root of several controversies in statistics. Even today, the heat of the debate has not decreased. Moreover, the theoretical gulf between Bayesianism and anti-Bayesianism is widening. A simple formulation of Bayes' theorem is the following. Let A and B be arbitrary events with probability $P(A)>0$ and $P(B)\geqq 0$, resp.; denote by $P(AB)$ the probability of the joint occurrence of A and B and by $P(A|B)$ the conditional probability of A if it is known that B has already been observed. Then

$$P(B|A)=\frac{P(AB)}{P(A)},\quad \text{i.e.,}\quad P(B|A)=\frac{P(A|B)P(B)}{P(A)}.$$

Therefore, if $B_0, B_1, \ldots$ are disjoint events with positive probability and one of them always occurs (or at least with probability 1) then

$$P(B_k|A)=\frac{P(A|B_k)P(B_k)}{P(A|B_0)P(B_0)+P(A|B_1)P(B_1)+\ldots}.$$

This is the Bayes' formula. It shows how *a priory* probabilities $P(B_k)$ (the probabilities of B_k before A was observed) determines the *a posteriori* probabilities (after A was observed). If the events B_k are considered the reasons, then Bayes' formula is a theorem on the probability of reasons. The theorem itself is indisputable but in most applications the probabilities $P(B_k)$ are unknown. In this case, it is typical, though generally not acceptable, to think that the absense of previous information on the reasons B_k implies the equality of the probabilities $P(B_k)$. Bayes applied his theorem in cases when *a priori* probabilities were of continuous distribution, especially when they were of uniform distribution on the interval (0, 1). According to the Bayes' theorem, if an event of unknown probability p occurs n times out of $n+m$ observations then the probability for p to belong to a subinterval (a, b) of the interval (0, 1) is

$$\frac{\int_a^b x^n(1-x)^m\,dx}{\int_0^1 x^n(1-x)^m\,dx}.$$

Bayes set out the idea that if we do not have any previous information about p then the *a priory* probability density of p is uniform on the whole interval (0, 1). If, e.g., $n=1, m=0, a=1/2$ and $b=1$, according to the above formula, the chance is 3/4 that the event in question has a probability more than 1/2. Still, few people would bet on the basis of this result partly because they doubt that the *a priori* distribution is uniform.

The lack of knowledge of *a priori* distributions had such a damaging effect on the statistical conclusions of Bayes' theorem that it has been almost excluded from the main line of statistics. In the second third of the 20th century, however, Bayesian conclusions were revived partly because of their essential role in finding admissible and minimax estimators (see I.12. Remarks and *Ferguson's* book). The opinion also gained ground that successive applications of the Bayes' formula (after each observation the *a posteriori* probabilities are calculated and used as *a priori* probabilities next time) reduce the importance of the original *a priori* distribution since after many repetitions the original distribution can hardly influence the final *a posteriori* distribution. (Obviously, certain degenerated cases are disregarded, e.g., when the value of p is 1/10, and the a priori distribution is uniformly distributed on the interval [1/2, 1] which does not cover the point 1/10.)

b) The paradox

Let the possible values of a random variable X be the integers and suppose that the probability distribution of X depends on a paramter p belonging to an interval $[a, b]$. If independent observations $X_1, X_2, X_3, \dots$ are made on the unknown distribution of X (i.e., on the unknown parameter p of the distribution; X_i are of the same distribution as X) then one can expect that the series of *a posteriori* distributions (calculated from the originally uniform *a priori* distribution) concentrates more and more on the true value of p. Paradoxically, this is not always true. The true value of p may be, e.g., 1/4 but the series of *a posteriori* distributions (as more observations are made) concentrates more and more, e.g., on 3/4.

c) The explanation of the paradox

The paradox seems to be surprising because the *a posteriori* density function is expected to be the highest in the neighbourhood of the true value, i.e., around 1/4. This idea, however, does not contradict the fact that the *a posteriori* density functions can concentrate more and more around 3/4. What should be achieved is only that the density function which is too high at 1/4 should very quickly decrease but remain high around 3/4. If the number of possible values of X is finite then this situation cannot be achieved, whereas if X can take any integer number then the paradoxical situation may really occur. Let the a priori distribution of p be uniform over the interval [1/8, 7/8]. Now let us define a function $f(p)$ on this interval in such a way that $f(p)$ is always a natural number except if $p=1/4$ or $p=3/4$ where $f(1/4)=f(3/4)=+\infty$. Let the distribution of the random variable X (depending on p) be the following:

$$P(X=i) = c(1-p)\, p^i, \quad i = 0, 1, 2, \dots, f(p)$$

where $c = c_p$ is a constant for which

$$\sum_{i=1}^{f(p)} c(1-p)p^i = 1.$$

By a suitable choice of $f(p)$, the above mentioned paradoxical situation becomes achievable. For further details see *Freedman*'s paper.

d) Remarks

(*i*) *S. Bernstein* and *R. Mises* had already pointed out before 1920 that, under some conditions, when applying Bayes' theorem successively, the series of *a posteriori* distributions always converge to the actual distribution whatever the *a priori* distribution was. That is why the *a priori* distribution has no significance asymptotically. According to the paradox, this conclusion cannot hold without any condition.

(*ii*) The subjective selection of *a priori* distributions raise the general question of whether unknown probabilities and probability distributions are objectively determined at all, independently of our observations and measurements, or they make sense only through our subjective information. In his monograph *Bruno de Finetti,* the head of the Italian school of probability theory, expresses that probability does not exist objectively, just as absolute space and time, the cosmic ether, or the phlogiston do not either. "Objective probability" is nothing else than an attempt to exteriorize and materialize our probabilistic beliefs. In his opinion an event (e.g., tomorrow it will rain) either occurs or not (this is objective), and on the basis of the information available we can figure out its "subjective" probability. The personal or subjective probability indicates the ratio of the bet we are willing to pay on the occurrence of the event. We can speak about subjective probability even if "randomness" is not objective. It has to be underlined, however, that the group of scientists claiming the existence of objective randomness and objective probability is much larger. Their conviction is the following: the objective probabilities of future events are encoded in the present state of the world. This kind of objective existence of probability has been expressed by the Nobel prize winner *Max Born,* who is famous for introducing objective probability into quantum physics.

e) References

Bayes, T., "An essay towards solving a problem in the doctrine of chances", 1763, Reprint: *Biometrika*, **45**, 293—315, (1958).

Berkson, J., "My encounter with neo-Bayesianism", *Internat., Statist. Rev.*, **45**, 1—9, (1977).

Born, M., *Natural Philosophy of Cause and Chance*, Dover, New York, 1964.

David, A. P., Stone, M., Zidek, J. V., "Marginalization paradoxes in Bayesian and structural inference", *J. Roy Statist. Soc., Ser. B.*, **35**, 189—233, (1973).

Ferguson, T. S., *Mathematical Statistics, A Decision Theoretic Approach.*, Academic Press, New York—London, 1967.

de Finetti, B., *Theorie Delle Probabilità*, Einaudi, Torino, 1970.

de Finetti, B., "Bayesianism", *Internat. Statist. Rev.*, **42**, 117—130, (1974).

Freedman, D. F., "On the asymptotic behavior of Bayes' estimates in the discrete case", *Annals of Math., Statist.* **34**, 1386—1403, (1963).

Holland, J. D., "The reverend Thomas Bayes F.P.S. (1702—1761)", *J. Roy. Statist. Soc.* (A), **125**, 451—461, (1962).

Lindley, D. V., "The use of prior probability distributions in statistical inference and decision", *Proc. 4th Berkeley Symp. on Math. Statistist. and Prob.*, **1**, 453—468, (1960).

Lindely, D. V., "The future of statistics — a Bayesian 21th century", *Advances in Appl. Prob.*, 106—115, (1975).

Lindley, D. V., "A problem in forensic science", *Biometrika*, **64**, 207—213, (1977).

Lindley, D. V., "The Bayesian approach", *Scand. J. Statist.*, **5**, 1—26, (1978). (3rd point: Marginalization paradoxes)

Pearson, E. S. (ed.), *The History of Statistics in the 17th and 18th Centuries Against the Changing Background of Intellectual, Scientific and Religious Thought*, Lectures by K. Pearson given at University College London during the academic sessions 1921—1933. Griffin, London, 1978.

Pflug, G., *Decision Theoretic Paradoxes in Decision Making under Uncertainty*, (ed. R. W. Sholtz). Elsevier, 375—383, 1983.

Savage, L. J., *The foundations of Statistics*, Dover, New York, 1972.

Stone, M., Springer, B. G. F., "A paradox involving quasi prior distributions", *Biometrika*, **52**, 623—627, (1965).

Shafer, G., "Lindley's paradox", *J. Amer. Statist. Assoc.*, **77**, 325—334, (1982).

2. PARADOX ESTIMATORS OF THE EXPECTATION

a) The history of the paradox

Equalization of contrasts and deviations in the "mean", i.e., summarizing the observations into a single value has long traditions. *Aeschylus* writes in the *Eumenides:* "To moderation in every form God giveth the victory, but his other dispensation he directeth in varying wise...", and the followers of the Chinese philosopher *Confucius* said that "the immobility of the mean (=Chung Yung) is the greatest perfection". Mathematically, the notion of "mean" can be interpreted in many ways (arithmetical mean, geometrical mean, median, etc.). In the practice of statistics, however, arithmetical mean was extremely important for a very long time. The first outstanding results in probability theory and in mathematical statistics also explored and reinforced the importance of the arithmetical mean of statistical samples.

Consider a set $\{F_\vartheta\}$ $\vartheta\in\Theta$ of probability distributions with finite expectation, where the parameter ϑ is just the expected value of F_ϑ. We want to estimate the value of the unknown parameter ϑ on the basis of the observed data (i.e., sample), $X_1, X_2, \ldots, X_n$, (the sample elements X_i are supposed to be independent, F_ϑ distributed random variables). The arithmetical mean

$$\hat{\vartheta} = \overline{X} = \frac{X_1+X_2+\ldots+X_n}{n}$$

as the estimator of ϑ has many good properties, e.g., it is always unbiased, that is, $E(\hat{\vartheta})=\vartheta$ for all $\vartheta\in\Theta$ (i.e., the estimate fluctuates around the actual value). The laws of large numbers state that the estimator $\hat{\vartheta}=\overline{X}$ is consistent, i.e., for any $\varepsilon>0$ we have

$$\lim_{n\to\infty} P(|\hat{\vartheta}-\vartheta| < \varepsilon) = 1 \quad \text{for all} \quad \vartheta\in\Theta,$$

so the error of the estimate can be made as small as desired by taking a sufficiently large sample. Nevertheless there may exist many unbiased, consistent estimators of a parameter and it is useful to give preference to estimators (among these) which have smaller variance. The paradoxes here reveal that (except in the case of normal distributions) the arithmetical mean of the sample is not the minimum variance unbiased estimator of the expected value. Moreover, if we do not insist on unbiasedness, then, even in the case of multivariate normal distributions, it is not always useful to estimate its expected value by the sample mean, because this estimator is not admissible with respect to the quadratic loss-function. [For the definition of admissible estimation see I.12. Remark (*ii*).] A similar paradox will be discussed in 13/q.

b) The paradoxes

(*i*) *(Kagan—Linnik—Rao)* Let $F(x)$ be an arbitrary distribution function with zero expectation and finite standard deviation and let $F_\vartheta(x)= =F(x-\vartheta)$ where the parameter ϑ is an arbitrary real value. If the elements of the sample $X_1, X_2, \ldots, X_n$ are random variables with distribution F_ϑ, then the sapmle mean $\overline{X}$ is a consistent and unbiased estimator of the unknown parameter ϑ (which is obviously the expected value of the distribution F_ϑ). The estimator $\overline{X}$, however, is not very efficient (except in the case of normal distribution): for any $n>2$, there exist an unbiased estimator the standard deviation of which is smaller than that of $\overline{X}$ (to be more precise, for all ϑ its standard deviation is at least as small as that of $\overline{X}$ and for at least one ϑ it is definitely smaller).

(*ii*) *(C. Stein)* $\overline{X}$ is an "exemplary good" estimator of the expectation of normal distributions: it is a minimum variance unbiased, consistent estimator, admissible in respect of the quadratic loss-function $L(\vartheta, c)= =(\vartheta-c)^2$, and also minimax. This is exactly why, some 20 years ago, C. Stein's discovery—claiming that in the case of multivariate normal distributions the estimator corresponding to $\overline{X}$ is not admissible—came as a surprise. More specifically, consider probability distributions defined on the k-dimensional Euclidean space the coordinates of which are (for simplicity) independent normal distributions $N(\vartheta, \sigma)$, where the standard deviation σ is known. We seek an admissible estimator $\hat{\vartheta}$ of the vector $\vartheta=(\vartheta_1, \vartheta_2, \ldots, \vartheta_k)$ whose quadratic loss

$$L(\vartheta, \hat{\vartheta}) = \|\vartheta-\hat{\vartheta}\|^2 = \sum_{i=1}^{k} (\vartheta_i-\hat{\vartheta}_i)^2$$

is, on the average, minimal. Then the vector $\hat{\vartheta}=\overline{X}$ (the k-dimensional vector of the sample average) is admissible only in 1 and 2 dimensions but not in higher dimensions (although the minimax property of $\overline{X}$ remains valid). Stein's recognition shows that even if we consider the classical estimating problem (that is, estimating the expected value of a normal distribution), $\overline{X}$ is not the only estimator we have to take into account.

c) The explanation of the paradox

(*i*) The interesting result of Kagan, Linnik, and Rao calls for proof rather than explanation. Instead of reproducing the proof, here is a method for finding asymptotically optimal estimators. First of all consider the example of a uniform distribution function $F(x)$ on the interval $(-c, c)$, (where c is an arbitrary positive number), and let $F_\vartheta(x)=$ $=F(x-\vartheta)$; then $D^2(\overline{X})=c^2/3n$. If $X_1^*=\min X_i$ and $X_n^*=\max X_i$, i.e., X_1^* is the smallest and X_n^* the largest sample element (they are both uniquely defined with probability 1 since the distribution is continuous) then

$$D^2\left(\frac{X_1^*+X_n^*}{2}\right)=\frac{2c^2}{(n+1)(n+2)},$$

which is far smaller than $D^2(\overline{X})$ for large n. Since

$$\frac{X_1^*+X_2^*}{2}$$

is also a simple unbiased and consistent estimator of ϑ, it is preferable to the "customary" $\overline{X}$. Turning to the general case let $X_1^*\leqq X_2^*\leqq\ldots\leqq X_n^*$ be the *ordered sample* (i.e., X_1^* is the smallest from $X_1, X_2, \ldots, X_n$, etc.) and

$$\tilde{X}=\sum_{i=1}^{n} a_{in}X_i^*$$

where a_{in} $(i=1, 2, \ldots, n;\ n=1, 2, \ldots)$ are real numbers depending on F. One can show that under some mild conditions the following choice of a_{in} leads to a minimum variance unbiased estimator of ϑ (at least asymptotically as $n\to\infty$). Let $a(x)$ be a real valued function on $[0, 1]$ and

$a_{in}=a(i/n)/n$. If F is 3 times differentiable then the optimal choice of $a(x)$ is defined by

$$a(F(x)) = -[(A+Bx)(\log f(x))']'$$

where

$$A = \frac{\mu_2}{\mu_0\mu_2-\mu_1^2}, \quad B = \frac{\mu_1}{\mu_0\mu_2-\mu_1^2},$$

$$\mu_0 = \int \frac{f'(x)^2}{f(x)}\,dx, \quad \mu_1 = \int \frac{f'(x)^2}{f(x)}\,x\,dx$$

and

$$\mu_2 = \int \frac{f'(x)^2}{f(x)}\,x^2\,dx - 1$$

(the primes denote the derivatives, $F'=f$ denotes the density function). This formula for $a(x)$ can be applied even if the expectation of F does not exist and ϑ denotes the center of symmetry of $f_\vartheta(x)$. E.g., let $f_\vartheta(x)= = \frac{1}{\pi(1+(x-\vartheta)^2)}$ (the Cauchy density; see the history of II/4). Then, surprisingly enough, $a(x)=-A\cos 2\pi x \sin^2 \pi x$ is the optimal choice which is negative (!) when x is close to 0 or 1. In this case the "customary" $\overline{X}$ estimator is not even consistent. (For a detailed analysis of this topic see T. F. Móri—G. J. Székely, "How to estimate location and scale parameters", Technical report, Eötvös L. Univ. 1986, see also H. Chernoff, J. L., Gastwirth and M. V. Johns, Jr. "Asymptotic distribution of linear combinations of order statistics with applications to estimation", *Annals of Math. Statistics,* **38**, 52—72, (1967).)

(*ii*) After Stein's article, which was published in 1956, James and Stein suggested the following simple estimator for the expectation of a multivariate normal distribution in 1961:

$$X^* = \left(1-\frac{(k-2)\sigma^2}{\|\overline{X}\|^2}\right)\overline{X}, \quad \text{where} \quad k>2.$$

Then $E\|X^*-\vartheta\|^2<k$, whereas $E\|\overline{X}-\vartheta\|^2=k$, hence the estimator $\overline{X}$ is really not admissible. The estimator X^* contracts the vector $\overline{X}$ towards the origin of the coordinate system, and as the origin can be chosen

arbitrary, the estimator

$$Q+\left(1-\frac{(k-2)\sigma^2}{\|\bar{X}-Q\|^2}\right)(\bar{X}-Q)$$

is also better than $\bar{X}$ for any Q. Thus the James—Stein estimator depends on how we choose the origin Q, whereas $\bar{X}$ is independent of Q. (It can be shown that the estimator

$$\tilde{X} = \max\left\{1-\frac{(k-2)^2}{\|\bar{X}\|^2};\ 0\right\}\bar{X}$$

is even slightly better than X^*.)

Now we shall turn our attention to the heuristic explanation of why the estimator X^* is better than $\bar{X}$. Consider the samples of k independent estimator problems together. The dispersion of the scalar sample elements is due, partly, to a (common) standard deviation σ of the k distributions and partly to the (generally) unequal expectations ϑ_i. Although these unknown expectations may be quite different, the combined sample may still show a dispersion which indicates that the values ϑ_i actually do not differ considerably. For example, in the case where $\sigma=1$ and about 16% of the observations are greater than 1, and 16% of them are smaller than -1, it is reasonable to think that all the expectations ϑ_i are near to zero. In this case, if $X_i=0.8$, the usual estimate of the ith parameter is 0.8, whereas—according to the more "rational" conception of the James—Stein estimator—the expectation ϑ_i is nearly zero. Though this explanation may convince us of the "rationality" of the James—Stein type estimations, it still seems extraordinary if we want to estimate, for example the expected values of the (normally distributed) body height, velocity of light and that of the price of a product, there can be any kind of connection between these problems.

d) Remarks

(*i*) The following inequality due to *Cramér* and *Rao* gives useful information concerning both paradoxes. Let $f(x, \vartheta)$ be the common density function (depending on the parameter ϑ) of the sample elements X_i, $i=1, 2, \ldots, n$ and let $B(\vartheta)$ denote the bias $E(\hat{\vartheta}_n-\vartheta)$ of the estimator

$\hat{\vartheta}$ of ϑ. Then the Cramér—Rao inequality claims that

$$E(\hat{\vartheta}_n-\vartheta)^2 \geqq B(\vartheta)^2+\frac{(1+B'(\vartheta))^2}{nI(\vartheta)}$$

holds under certain regularity conditions, where the function $B'(\vartheta)$ is the derivative of $B(\vartheta)$ and

$$I(\vartheta) = E\frac{-d^2 \ln f(X_i, \vartheta)}{d^2\vartheta}$$

is the Fisher information. Thus the rate of convergence of $E(\hat{\vartheta}_n-\vartheta)^2$ to zero cannot exceed $1/n$. However, in the example of the first paradox (uniform distribution, $B(\vartheta)\equiv 0$ and therefore $E(\hat{\vartheta}_n-\vartheta)^2=D^2(\hat{\vartheta}_n)$) the rate of convergence is $1/n^2$. This is not a contradiction because the example in question is a typical case where the above mentioned regularity conditions do not hold. (The following condition, for example, would be a sufficient regularity condition: the set of the numbers x where $f(x, \vartheta)$ is positive does not depend on ϑ.) Concerning our second paradox, the Cramér—Rao inequality points out that allowing biased estimators, i.e., if we drop the condition $B(\vartheta)\equiv 0$, and if $B'(\vartheta)$ happens to be negative, $E(\hat{\vartheta}_n-\vartheta)^2$ might decrease considerably compared to the variance of the minimum variance unbiased estimator.

(*iii*) Let $X_1^*<X_2^*<\ldots<X_n^*$ denote an ordered sample and let X' denote the sample median, i.e., $X'=X^*_{(n+1)/2}$ if n is odd and

$$X' = \frac{X^*_{n/2}+X^*_{n/2+1}}{2}$$

if n is even. If we take the sample from a normal distribution then

$$D^2(\overline{X}) \approx \frac{2}{\pi} D^2(X') \approx 0.63 D^2(X'),$$

i.e., the efficiency of the estimator X' is (asymptotically) only 63% of that of $\overline{X}$. The situation changes, however, if we slightly "perturb" the normal distribution: consider a random variable which is the mixture of two normal distributions, namely 91% is a normal distribution with expectation ϑ and variance 1, and 9% is a normal distribution with the same expectation ϑ and variance 9. In this case the median X' is a better estimator of ϑ than $\overline{X}$.

(*ii*) The following paradox of admissible estimate is due to S. M. Masani. Let X_1, X_2 be two independent random variables with expectations m_1, m_2. Masani gives two examples (one binomial, one normal) in which an estimator, depending only on X_2, is admissible for estimating m_1. Here we mention only the binomial case. Let $X_1, X_2, \ldots, X_m$ be independent binomial random variables with parameters n_i, p_i. One can prove that a necessary and sufficient condition for the linear estimator

$$P_1(X_1, X_2, \ldots, X_m) = \sum_{i=1}^{m} a_i X_i/n_i + c$$

to be admissible for p_1, when the loss function is the quadratic loss $L(p_1, p) = (p_1 - p)^2$, is that either

$$0 \leqq a_1 < 1, \quad 0 \leqq c \leqq 1$$

and

$$0 \leqq \sum_{i=2}^{m} a_i + c \leqq 1, \quad 0 \leqq \sum_{i=1}^{m} a_i + c \leqq 1$$

or $a_1 = 1$ and $a_2 = a_3 = \ldots = a_m = c = 0$. If we put $a_1 = 0$ then we get a large class of admissible estimators for p_1 not depending on X_1.

e) References

Efron, B., "Biased versus unbiased estimation", *Advances in Math.,* **16**, 259—277, (1975).

James, W., Stein, C., "Estimation with quadratic loss", *Proc. 4th Berkeley Symp. on Math. Statist. and Prob.,* **1**, 361—380. Univ. California Press, Berkeley, (1961).

Kagan, A. M., Linnik, Yu. V., Rao, C. R., "On a characterization of the normal law based on a property of the sample average", *Sankhya,* Ser. A, **27**, 3—4, 405—406, (1965).

Masani, S. M., "A paradox in admissibility", *Annals of Statist.,* **5**, 544—546, (1977).

Stein, C., "Inadmissibility of the usual estimator for the mean of a multivariate normal distribution", *Proc. 3rd Berkeley Symp. on Math. Statist. and Prob.,* **1**, 197—206, Univ. California Press, Berkeley, (1956).

Tukey, J. W., "A survey of sampling from contaminated distributions", Contrib. to Prob. and Statist, (Ed. I. Olkin) 448—485), Stanford Univ. Press, 1960.

Zacks, S., *The Theory of Statistical Inference,* Wiley, New York, 1971.

3. PARADOX ESTIMATORS OF THE VARIANCE

a) The history of the paradox

Besides expected value, variance is the other most important characteristic of random variables and their distributions. Estimate the unknown variance D^2 of the random variable X from the sample $X_1, X_2, \ldots, X_n$ (these are independent observations and have the same distribution as X). If the expected value E is known then the estimator

$$\hat{D}_0^2 = \frac{1}{n} \sum_{i=1}^{n} (X_i - E)^2$$

is unbiased. The situation changes if E is unknown and is replaced (in the above formula) by its unbiased estimator $\overline{X}$. Then the estimator

$$\hat{D}^2 = \frac{1}{n} \sum_{i=1}^{n} (X_i - \overline{X})^2$$

is no longer unbiased. Since unbiasedness has been one of the most required good properties of estimators (since Gauss' time) the estimator $\hat{D}^2$ was modified to make it unbiased. (Several parameters do not have unbiased estimators at all. In these cases only *asymptotic unbiasedness*, i.e.,

$$\lim_{n \to \infty} E(\hat{\vartheta}_n) = \vartheta$$

for all $\vartheta \in \Theta$ is required. This property holds for $\hat{D}^2$.) Besides unbiasedness other important properties of good estimators were crystallized. A paradox appears when different properties of good estimators do not lead to the same estimator.

b) The paradox

Multiplying $\hat{D}^2$ by the *Bessel* factor $\dfrac{n}{n-1}$ we get

$$D^{*2} = \frac{1}{n-1} \sum_{i=1}^{n} (X_i - \overline{X})^2$$

which is an unbiased estimator of the variance.

Suppose that X is normally distributed (with unknown expectation and variance) and we prefer minimax estimators (see I.12. Remarks) with loss function $L(\tilde{D}^2, D^2)=(\tilde{D}^2-D^2)/D^4$; then $\tilde{D}^2$ must be modified in just the other direction: it has to be multiplied by $\frac{n}{n+1}$ to obtain a minimax estimator:

$$\frac{1}{n+1}\sum_{i=1}^{n}(X_i-\overline{X})^2.$$

$\left(\text{Its risk is } \frac{2}{n+1}.\right)$ Thus the minimax principle and unbiasedness led to different estimators.

c) The explanation of the paradox

The sum

$$\sum_{i=1}^{n}(X_i-a)^2$$

is minimal only if $a=\overline{X}$. However, the expected value E is generally not equal to $\overline{X}$ (only near it), therefore $\hat{D}_0^2$ showing the real deviation is greater than $\hat{D}^2$. That is why the Bessel correction is needed. On the other hand there is no reason why minimax or admissible estimators should be unbiased. (We have already seen in II.2. that the James—Stein estimator of the expected value is better than the usual unbiased estimator $\overline{X}$.) Since the unbiased and minimax estimators of the variance of normal distributions do not coincide, we have to decide with each practical problem which one to choose. Fortunately, there is only a slight difference between the two estimators even at small values of n. (In other problems the difference, however, can be significant.)

d) Remarks

Though surprising, it can be proved that the above mentioned minimax estimator is not admissible. (See Stein's article or Zacks' book.) On the other hand, if the expected value of the normal distribution is known, the estimator

$$\frac{1}{n+2}\sum_{i=1}^{n}(X_i-E)^2$$

is not only minimax $\left(\text{with a risk of } \frac{2}{n+2}\right)$ but also admissible regarding the mentioned loss function. (See Zack's book.)

e) References

Stein, C., "Inadmissibility of the usual estimate for the variance of a normal distribution with unknown mean", *Annals Inst. Statist. Math.*, **16**, 155—160, (1964).
Zacks, S., *The Theory of Statistical Inference*, Wiley, New York—London—Sydney—Toronto, 1971.

4. THE PARADOX OF LEAST SQUARES

a) The history of the paradox

Due to the inevitable errors of observed measurements, theoretical formulas and empirical data frequently seem to be in contradiction. *Legendre, Gauss*, and *Laplace* elaborated an efficient method to diminish the effect of measurement errors early in the last century. (Legendre, for instance, worked out and applied it in 1805 to determine the orbits of comets.) The pioneers of this theory were *Galileo* (1632), *Lambert* (1760), *Euler* (1778), and others. The new procedure, called the method of least squares was discussed in detail by Gauss in his work "Theoria Motus" (1809). It was also Gauss who pointed out the probabilistic background of the method. (Though Legendre accused Gauss of plagiarism, Legendre could not properly substantiate his repeated accusations. Gauss claimed priority only in the use of the method and not in its publication.) Laplace published his fundamental book on probability theory in 1812 which he dedicated to "Napoleon the Great". The entire fourth chapter is devoted to the calculus of error. Since then the method of least squares has developed into a new branch of mathematics. It is sometimes "overmystified" and often used when other methods would be more expedient.

This problem was emphasized even by Cauchy (Comptes Rendus, 1853) during his "debate" with *Bienaymé* (in the course of this dispute Cauchy used the probability density function $1/\pi(1+x^2)$, which was later named after him, though he was not the first scientist to use the "Cauchy density").

b) The paradox

Let $ae^{-b|x-\mu|}$ be the density function of our observations subject to random measurement errors; the constants a and b are known and μ has to be estimated. We make independent observations $X_1, X_2, \ldots, X_n$. According to the method of least squares, μ has to be estimated by the value $\hat{\mu}$ which minimizes the sum

$$(X_i-\hat{\mu})^2+(X_2-\hat{\mu})^2+\ldots+(X_n-\hat{\mu})^2.$$

It is easy to calculate that this sum is minimal if $\hat{\mu}$ is the arithmetical mean of the observed data:

$$\hat{\mu} = \frac{X_1+X_2+\ldots+X_n}{n}.$$

However, if we prefer the estimator $\bar{\mu}$ for which the probability (more precisely the probability density) that the results of n observations are just $X_1, X_2, \ldots, X_n$ is maximal, i.e., if $\bar{\mu}$ maximizes

$$a^n e^{-b(|X_1-\mu|+\ldots+|X_n-\mu|)}$$

or, equivalently, $\bar{\mu}$ minimizes

$$|X_1-\bar{\mu}|+|X_2-\bar{\mu}|+\ldots+|X_n-\bar{\mu}|,$$

then we get a contradiction since the sum of squares and the sum of absolute values do not take their minimum at the same value of μ, i.e., $\hat{\mu}$ and $\bar{\mu}$ are different. Which one is better?

c) The explanation of the paradox

If the measurement errors were normally distributed (i.e., if their density function were of the form

$$ae^{-b(x-\mu)2},$$

the above mentioned contradiction would not appear since $\hat{\mu}$ maximizes

$$a^n e^{-b((X_1-\mu)^2+\ldots+(X_n-\mu)^2)}.$$

Gauss based the method of least squares on normally distributed errors, and in practice this is the most frequent case. However, if the distribution of errors is known to be different from normal, then using the least squares estimator is not always advantageous. In the case of the above mentioned paradox, the use of the estimator $\bar{\mu}$ is more reasonable (see also the previous section).

Using the customary notions of mathematical statistics, the paradox can be formulated in brief as follows: the least squares estimator is not always compatible with the maximum-likelihood estimator (for maximum likelihood estimation see Section 8). In fact if $f(x)$ is a positive semi-continuous density function at $x=0$ (from below), and the density of the observations is $f(x-\vartheta)$ and

$$\bar{X} = \frac{X_1+X_2+\ldots+X_n}{n}$$

is a maximum likelihood estimator of ϑ for $n=2,3$, then $f(x)$ is the density function of a normal distribution with zero expectation. This remark is the Gaussian law of error, and it can be proved as follows: if for simplicity the existence of the derivative f' is supposed and

$$\prod_{i=1}^{n} f(X_i-\vartheta)$$

is maximal for $\vartheta=\bar{X}$ then

$$\sum_{i=1}^{n} \frac{f'}{f}(X_i-\bar{X}) = 0,$$

i.e., (with the notation $\Delta_i=X_i-\bar{X}$)

$$\sum_{i=1}^{n} \Delta_i = 0 \quad \text{implies} \quad \sum_{i=1}^{n} \frac{f'}{f}(\Delta_i) = 0$$

and this can be valied for $n=2,3$ (when f'/f is measurable) only if $\frac{f'}{f}(x)=cx$, and it follows that $f=de^{-cx^2}$ where c and d are positive numbers (otherwise f would not be a density function). So the least square estimator of the location parameter can coincide with its maximum likelihood estimator only for normal distributions.

d) Remark

The arithmetic mean $\hat{\mu}=\bar{X}$ and the median $\tilde{\mu}$ are the only "simple" maximum likelihood estimators of the parameter μ having the form

$$L = \sum_{i=1}^{n} a_i X_i^*$$

where $X_1^* \leqq X_2^* \leqq \ldots \leqq X_n^*$ is the ordered sample and $\sum_{i=1}^{n} a_i = 1$ $(n=1, 2, \ldots)$.

e) References

Berkson, J., "Estimation by least squares and by maximum likelihood", *Proc. 3rd Berkeley Symp. on Math. Statist. and Prob.*, I, 1—11, (1956).
Bloomfield, P. B., Steiger, W. L., *Applications of Least Absolute Deviations*, Birkhäuser Verlag, Basel—Boston—Stuttgart, 1983.
Harter, H. L., "The method of least squares and some alternatives", Part. I—V. *Internat. Statist. Rev.* (1974—1975).
Linnik, Yu. V., *Die Methode der kleinsten Quadrate in moderner Darstellung*, Deutscher Verl. der Wiss., Berlin, 1961.
Sheynin, O. B., "C. F. Gauss and the theory of errors", *Archive for History of Exact Sciences*, **19**, 21—72, (1979).
Stigler, S. M., "Cauchy and the witch of Agnesi", *Biometrika*, **61**, 375—380, (1974)

5. CORRELATION PARADOXES

a) The history of the paradox

By the last third of the previous century, several sciences (e.g., molecular physics) had reached such a level of development that the adaptation of probability theory and mathematical statistics became indispensable in these fields too. In 1859 *Darwin*'s book revolutionized biology, and shortly afterwards his cousin *Francis Galton* established human genetics. (*Mendel*'s study on genetics was only "rediscovered" at the turn of the century, and the word genetics has only been used since 1905; but Galton's results had already aroused great interest in the last century.) Galton and his students (especially *Karl Pearson*) introduced many important notions such as *correlation* and *regression* which became fundamental ideas of both probability theory and mathematical statistics (as well as of other related sciences). A man's weight and height are, naturally, in close connection though they do not determine each other uniquely. Correlation measures this connection by a single number the absolute value of which is not more than 1. The correlation of two random values X and Y is defined as follows. Let E_x and D_x, E_y and D_y denote the expected value and the standard deviation of X and Y, resp. Then the correlation coefficient (in brief: correlation) of X and Y is

$$r = r(X, Y) = \frac{E[(X-E_x)(Y-E_y)]}{D_x D_y}.$$

The absolute value of the correlation is maximal (i.e., $=1$) if there is a linear relationship between X and Y, i.e., $Y=aX+b$ (where $a\neq 0$). If X and Y are independent (and their variance is finite) then their correlation is 0, i.e., they are uncorrelated. In mathematical statistics the correlation r is usually estimated from the generally independent sample $(X_1, Y_1), (X_2, Y_2), \ldots, (X_n, Y_n)$ by the following sample correlation coefficient

$$\hat{r} = \frac{\sum_{i=1}^{n} (X_i-\bar{X})(Y_i-\bar{Y})}{\sqrt{\sum_{i=1}^{n} (X_i-\bar{X})^2 \sum_{i=1}^{n} (Y_i-\bar{Y})^2}}.$$

In several cases r gives a good characterization of the relationship between X and Y but even at the turn of the century several senseless correlations were calculated, e.g., the correlation of the number of stork nests and that of infants. Correlations have gradually been mistificated and several "internal", generally casual, connections were thought to exist in the case of close (near to 1 in absolute value) correlation. This is why totally absurd results were created which nearly succeeded in discrediting statistics as a whole. It generally was ignored that close correlation of X and Y might be caused by a third quantity. E.g., it was observed in England and Wales that when the number of radio licences was increased there was a corresponding increase in the number of insane and mentally handicapped people. This interpretation is, however, completely false because listening to the radio does not bring about mental illness; it is simply that as time passes the number of radio listeners as well as the number of mental cases increase though there is no causal connection between them whatsoever. Unfortunately, misinterpretations are not always so simple to discover, e.g., in technical or economical applications. The comparison of religion and height is another senseless correlation, which claims that going from Scotland in the direction of Sicily the rate of Roman-Catholics gradually increases while the average height of people decreases gradually; but of course any causal interpretation is absolute nonsense. (Even more farcical ideas were claimed to be causal relationships and even science by Fascist racial theory.) We will mention only some of the existing correlation paradoxes.

b) The paradoxes

(*i*) Let X be uniformly distributed over the interval $(-1, 1)$ and $Y=|X|$. Obviously, there is a very close relationship between X and Y, but their correlation $r(X, Y)=0$. (The correlation of X and $Y=|X|$ is always 0 when X is a random variable with finite variance and has a symmetric distribution around 0.)

(*ii*) Let $X_1, X_2, \ldots, X_n$ be the temperature of a room in n different moments and $Y_1, Y_2, \ldots, Y_n$ be the quantity of the fuel used up for heating in the same moments (more precisely during a given period, e.g., 1 hour before these moments). It is logical to think that the more fuel used the warmer the room will be. It means that the correlation of X and Y is strictly positive. In spite of this the correlation may be negative, which can be interpreted as the more we heat the colder it will be.

(*iii*) Let (X, Y) be normally distributed, i.e., let the density function of (X, Y) be

$$f(x, y) =$$

$$=\frac{1}{2\pi D_x D_y \sqrt{1-r^2}} e^{\frac{1}{2(1-r^2)}\left[\left(\frac{X-E_x}{D_x}\right)^2-\frac{2r(X-E_x)(Y-E_y)}{D_x D_y}+\left(\frac{Y-E_y}{D_y}\right)^2\right]}$$

where E_x, D_x, E_y and D_y are the expected values and variances of X and Y, and r is their correlation. Now we suppose that the absolute value of the correlation is strictly less than 1. If r is unknown we can estimate it by $\hat{r}$ from a sample of n elements. If E_x and E_y are known then it is advisable to modify the formula of $\hat{r}$ so that $\bar{X}$ and $\bar{Y}$ are replaced by E_x and E_y, resp. In this way we obtain a new estimate $\bar{r}$. As $\bar{r}$ uses more information (namely the knowledge of E_x and E_y) we might think that its variance is less than that of $\hat{r}$. However, *A. Stuart* calculated that

$$D^2(\hat{r}) = \frac{1}{n}(1-r^2)^2 \quad \text{while} \quad D^2(\bar{r}) = \frac{1}{n}(1+r^2)$$

consequently the latter is the bigger.

c) The explanation of the paradoxes

(*i*) If X and Y are independent then $r(X, Y)=0$ but the inverse assertion is false. Uncorrelated values may be strongly dependent as in the above example where $Y=|X|$. Therefore "being uncorrelated" must not be interpreted simply as being independent. On the other hand, it can be proved that if X and Y are uncorrelated under the restrictions $x_1<X<x_2$, $y_1<Y<y_2$ whatever the number $x_1<x_2$ and $y_1<y_2$ be, X and Y are independent.

(*ii*) We must not forget the disturbing effect of the temperature outside!

We often obtain completely unbelievable correlations because the expected correlation coefficient of two random variables had been twisted by a third, "exterior disturbing variable". It is precisely to avoid such distrubing effects that the notion of *partial correlation* has been introduced. If the correlation of X and Y is calculated only after having filtered out the disturbing effect of the variable Z then the result will no longer be a paradox. Let r_{12}, r_{13} and r_{23} denote the correlations $r(X, Y)$, $r(X, Z)$ and $r(Y, Z)$, resp. Then the partical correlation of X and Y after having filtered out the effect of Z is

$$\frac{r_{12}-r_{13}r_{23}}{\sqrt{(1-r_{13}^2)(1-r_{23}^2)}}.$$

In the special case when $r_{13}=r_{23}=0$, the partial correlation of X and Y is equal to the simple correlation r_{12}. If r_{12}, r_{13}, r_{23} are not known then they can be estimated from the sample just like r. By the help of these estimators, we shall obtain an estimator for the partial correlation coefficient.

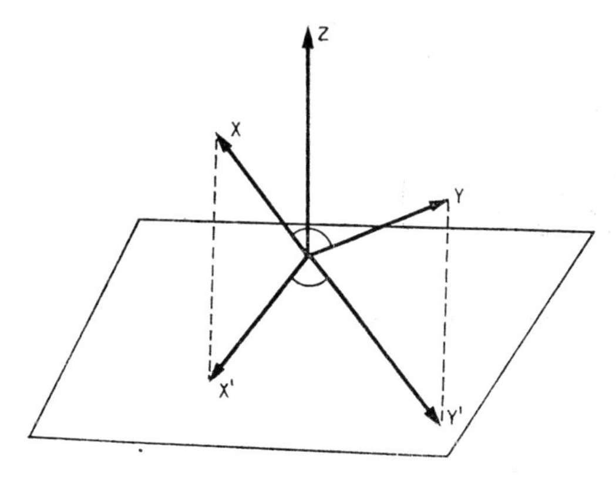

Figure 9. Considering the random variables X, Y, and Z as vectors, the correlation of random variables X and Y is the cosine of the angle between the vectors X and Y, and their partial correlation is the cosine of the angle between their images under projection onto the plane perpendicular to vector Z.

(*iii*) Stuart's paradox can be shown from many aspects. The main point is that $\hat{r}$ and $\bar{r}$ are not unbiased estimators of r, i.e., the identities $E(\hat{r})\equiv r$ and $E(\bar{r})\equiv r$ are not valid, and if so, it is not advisable to consider an estimator better if its variance is less. At the same time neither $\hat{r}$ nor $\bar{r}$ is very biased (asymptotically unbiased), therefore the explanation of the paradox needs further analysis. (See the Remarks and Stuart's article.)

d) Remarks

(*i*) The bias of the estimator r (in case of bivariate normal distribution) is the following

$$E(\hat{r}-r) = \frac{r^2-1}{n} + o(n^{-1})$$

where $o(n^{-1})$ denotes an expression which converges to 0 even if multiplied by n. Thus bias converges rather fast to 0 (as the sample size n increases). On the other hand, it is interesting that arcsin $\hat{r}$ is an unbiased estimate of arcsin r, and if $E(g(\hat{r}))=E(g(r))$ for some function g independent of n then $g(r)=a\cdot\arcsin r+b$, where a, b are arbitrary constants. In 1958 *I. Olkin* and *J. W. Pratt* proved: if the estimator of the correlation coefficient r may directly depend on n then one can find an unbiased estimator for r itself, namely

$$r^* = \hat{r}F\left(\frac{1}{2}, \frac{1}{2}, \frac{n-2}{2}, 1-\hat{r}^2\right)$$

where F is the hypergeometric function given by

$$F(x, a, b, c) =$$

$$= 1+\sum_{k=1}^{\infty} \frac{a(a+1)\dots(a+k-1)\,b(b+1)\dots(b+k-1)}{k!\,c(c+1)\dots(c+k-1)}\,x^k$$

where a, b, c ($c\neq 0, -1, -2, \dots$) are parameters. Among unbiased estimators it is already worth preferring those of minimal variance. It can be shown that r^* is not only unbiased but of minimal variance, too. However, r^* is rather complicated in practice, therefore it is advisable to apply the following approximation of it:

$$\hat{r}\left(1+\frac{1-\hat{r}^2}{2(n-4)}\right).$$

(*ii*) It is not a paradox but it is nevertheless surprising that in choosing m numbers randomly from $1, 2, \dots, n$ (sampling without replacement, i.e., the number of equally probable choices is $\binom{n}{m}$), the correlation coefficient of the smallest and greatest number chosen is $\frac{1}{m}$, i.e., it is independent of n ($m=1, 2, \dots, n-1$). More generally, if $X_1, X_2, \dots, X_m$

denote the increasing series of the m values selected, then

$$r(X_i, X_j) = \sqrt{\frac{i(m+1-j)}{j(m+1-i)}}$$

which is independent of n, too.

[The following result is due to T. F. Móri. Take a sample of n elements from $1, 2, \ldots, m+1$ with replacement and denote by Y_i the number of sample elements the value of which is not greater than i. Then $r(Y_i, Y_j) = r(X_i, X_j)$.] Another related problem is the following. If $Z_1 \leqq Z_2 \leqq \ldots \leqq Z_m$ is an increasing sequence of independent uniformly distributed random variables ("ordered sample") then $r(Z_i, Z_j) = r(X_i, X_j)$; if the uniform distribution is replaced by any other distribution for which $r(Z_i, Z_j)$ exists then $r(Z_i, Z_j) \leqq r(X_i, X_j)$, i.e., this is an extremal property of uniform distributions (see the paper by Móri—Székely). In fact we can prove more. Let the maximal correlation of two random variables U and V be defined as the supremum of $r(f(U), g(V))$ where f and g run over the set of square integrable real functions of U and V, resp.:

$$\max \operatorname{corr} (U, V) = \sup_{f,g} r(f(U), g(V)).$$

Using this notion, we can prove that

$$\max \operatorname{corr} (Z_i, Z_j) = r(X_i, X_j).$$

This follows from the fact that max corr $(U, V) = r(U, V)$ if both the regression of U on V (for the definition of regression see the next paradox) and the regression of V on U is linear (and not identically constant). It is precisely this case when correlation is a good measure of closeness.

e) References

Kendall, M. G., Stuart, A., *The Advanced Theory of Statistics*, Vol. 2. Griffin, London, 1961.

Móri, T. F., Székely, G. J., "An extremal property of rectangular distributions", *Statistics and Probability Letters* **3**, 107—109, 1985.

Olkin, I., Pratt, J. W., "Unbiased estimation of certain correlation coefficients", *Annals of Math. Statist.*, **29**, 201—211, (1958).

Stuart, A., "A paradox in statistical estimation", *Biometrika*, **42**, 527—529, (1955).

6. REGRESSION PARADOXES

a) The history of the paradoxes

The correlation coefficient measures the dependence between two random variables by a single number, whereas regression expresses this dependence by a function-like relation and thus gives more precise information. For example, the average body weights as a function of body heights is a regression. The name "regression" comes from *Galton,* who, at the end of the past century, compared the heights of parents with that of their children. He found that the heights of children with tall (or small) parents are usually above (or below) the average but not as much as their parents' heights. The line which showed to what extent the

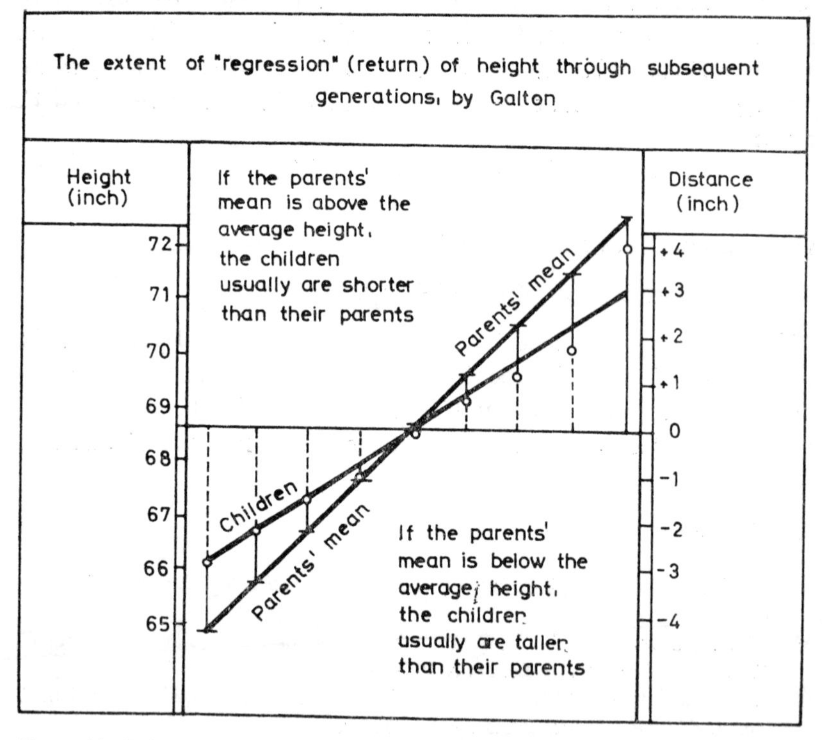

Figure 10. Galton's regression line.

heights (and other properties as it later turned out) regressed (returned) to the average through subsequent generations was called a *regression line* by Galton. Later any function-like relation between random variables were called regression. Regression analysis was applied first in biology, and the most important scientific journal which dealt with this topic was the *Biometrika,* published since October 1901. In the years between 1920 and 1930, its economic applications also became very important and a new branch of science sprang up: econometrics (the term is due to *R. Frisch,* 1926, who was later awarded the Nobel prize), with its own journal, *Econometrica,* first published in 1933. From examining special regression problems, researchers gradually came to the regression analysis of the intrinsic structure of global economic systems (*J. M. Keynes, J. Tinbergen* and others as, e.g., *R. L. Klein,* who won the Nobel prize for economics in 1980). The journal *Technometrics* has been published since 1959 and mainly deals with technical applications. The regression analysis of a quantity X on another quantity Y—where X is difficult to measure and Y can be measured quite easily—is very important Nowadays almost every branch of science applies regression analysis, which is a good thing in itself, but unfortunately regression analysis has also become one of the chief means of "facile scientific successes", slipshod analyses and glossing over (scientific) problems. Regression never substitutes scientific conceptions and theoretical background, though it can help to find them.

b) The paradoxes

Suppose the dependence of two variables is given by a function of the following type:

$$y = f(x; a_1, a_2, \dots, a_m), \quad (\text{e.g. } y = a_1 x + a_2),$$

where only the parameters $a_1, a_2, \dots, a_m$ are unknown, (the type of function, e.g., linear, quadratic, etc. is known). If we can measure the values of y only with random observational errors, i.e., instead of $y_i = f(x_i; a_1, a_2, \dots, a_m)$ we observe the values Y_i subject to errors, then, according to the method of least squares, the unknown a_i's minimize the sum of squares

$$\sum_{i=1}^{n} (Y_i - f(x_i;\ a_1, a_2, \ldots, a_m))^2.$$

(*i*) Accordingly, if $f(x) = e^{ax}$ then the estimator of a minimizes

$$\sum_{i=1}^{n} (Y_i - e^{ax_i})^2.$$

In this case the problem of calculating the regression-curve f is usually simplified by taking the logarithm of both terms of the difference in brackets and minimizing the quantity

$$\sum_{i=1}^{n} (\ln Y_i - ax_i)^2,$$

which can be easily performed by finding the minimum of a quadratic polinomial. However, the two minimizing methods give different estimators. What is the solution of this paradoxical situation?

(*ii*) Suppose we have only an alternative idea about the type of f, for example, f_1 is a polinomial and f_2 is an exponential function. It seems natural to accept the type for which the above sum of squares is smaller (under optimal choice of the parameters). Though this principle is often followed in practice, usually it is not reasonable (sometimes even the theoretical possibility of this choice must be questioned).

(*iii*) Let $y = ax$ be the theoretical regression line and let $Y_i = ax_i + \varepsilon_i$, where ε_i $(i = 1, 2, \ldots, n)$ are independent normally distributed random errors with expectation 0 and variance $D^2(\varepsilon_i) = cy$ (c is a known constant). Now suppose that the observations happen to fit the regression line perfectly, i.e., $Y_i = a_0 x_i$ for some a_0, thus

$$\sum_{i=1}^{n} (Y_i - a_0 x_i)^2 = 0.$$

Then the least squares estimator of a is a_0, but, paradoxically, this is not the "best" estimator (in the sense of maximum-likelihood; cf. definition in paradox 8).

c) The explanation of the paradoxes

(*i*) Undoubtedly the method of least squares corresponds to the first method, nevertheless it is useful to consider not only the letter but also the spirit of this principle, since the qualitative meaning of least squares is the minimization of the total effect of errors. This purpose can also be achieved by minimizing the sum of squares

$$\sum_{i=1}^{n} (h(Y_i)-h(f(x_i;\ a_1, a_2, \dots, a_m)))^2,$$

where $h(x)$ is a monotone increasing function (e.g., $h(x)=\ln x$). A good choice of h "linearizes", i.e., makes the formula $h(f(x_1;\ a_1, a_2, \dots$ $\dots, a_m))$ a linear function of the unknown parameters a_i (in this case the optimal value of a_i's can be easily determined). If we want to determine the unknown parameters in the spirit of least squares principle, it is clearly better to choose the second method. However, the original sum of squares, for example, may have to be minimized if we know that the errors result in a financial loss which is proportional to this sum of squares, though this possibility is far from typical.

(*ii*) The first part of the question is very simple: the sum of squares may be smaller for f_1 than for f_2, but, taking some more sample elements into consideration, the sum of squares becomes smaller if we choose f_2. Mathematical statistics tries to avoid such unstable situations. There are some decision methods available in certain cases, which decide with given, e.g., 99% certainty, (i.e., if f_1 is rejected then the probability that f_1 is the right choice is 1%). In *Plackett*'s book, for instance, a method is discussed which enables the proper degree of regression polynomials to be chosen (in case of independent, normally distributed observetional errors). Unfortunately, many typical alternative regression problems cannot be properly handled. For example, the *Weber—Fechner rule* states that there is a logarithmic relation between stimulus and sensation, especially between volume and sound intensity, or between frequency and pitch. Nowadays this rule is considered only an approximate one both theoretically and experimentally, because a power-function relation seems to be closer to the truth. (In fact, the problem is more complicated as the sensation of loudness depends not only on intensity but also on frequency and the spectrum of sound as well as on the duration of the experiment.)

(*iii*) The estimator $\hat{a}=a_0$ is not satisfactory since the estimator of $D^2(\varepsilon_i)$ would than be zero which contradicts the condition $D^2(\varepsilon_i)=cy$. The estimator

$$[(\sqrt{1+4c^2}-1)/(2c^2)]\, a_0$$

is more reasonable (maximum-likelihood).

d) Remarks

(*i*) The logit-probit alternative is also very typical—especially in pharmacology and market research. In logit analysis, the function

$$Y=\frac{e^{a_1+a_2x}}{1+e^{a_1+a_2x}}$$

is fitted to the data by the method of least squares, minimizing

$$\sum_{i=1}^{n}\left(\ln\frac{Y_i}{1-Y_i}-a_1-a_2x_i\right)^2.$$

[Here the transformation function which linearizes the problem is $h(x)=\ln\frac{x}{1-x}$.] In probit analysis a normal distribution function is fitted to the data (by an appropriate choice of the parameters). The shapes of the two types of curves may be quite similar, so it is not always easy to decide which one to choose, but the theoretical background may be a great help.

(*ii*) If we increase the number of regression parameters then obviously we obtain a better fitting but then the variances of parameter estimators increase, so the estimators become less stable and less reliable.

(*iii*) For the "paradox of the two regressions" see Kalman (1982). In this paper (following the pioneering work by Gini (1921) and Frisch (1934)) it is assumed that there are random (additive) errors in both variables: $X=x+\hat{x}$ and $Y=y+\hat{y}$ ($\hat{x}, \hat{y}$ are the errors or "noise"). Supposing $y=ax$, the "unprejudice estimate" of a can be given only in terms of an interval: $a_1\leqq a\leqq a_2$. Here one of the limits is the classical regression coefficient (when y is regressed on x), and the other limit is the reciprocal regression coefficient (when x is regressed on y). The choice of either limits a_1 or a_2 implies the prejudice that the regressor variable is noise-free. (This gives a solution for the "paradox of the two regressions".)

e) References

Berkson, J., "Minimum chi-square, not maximum likelihood!" *Annals of Stotist.*, **8**, 457—487, (1980).

Box, G. E. P., "Use and abuse of regression", *Technometrics*, **8**, (1966).

Box, G. E. P., Cox, D. R., "An analysis of transformations", *J. R. Statist. Soc. Ser. B*, **26**, 211—243, (1964).

Cox, D. R., *The Analysis of Binary Data*, Methuen, London, 1970.

Daniel, C., Wood, F. S., *Fitting Equation to Data*, Wiley, New York, 1971.

Draper, N. R., Smith, H., *Applied Regression Analysis*, Wiley, New York, 1966.

Durbin, J., "Errors in variables", *Rev. Inst. Int. Statist*, **22**, 23—32, (1954).

Frisch, R., "Statistical confluence analysis by means of complete regression systems", *Publ. No. 5, Univ. Oslo Economic Inst.*, 192 pages, (1934).

Gini, C., "Sull'interpolazione de una retta quando i valori della variabile indipendente sono affetti da errori accidentali", *Metron*, **1**, 63—82, (1921).

Kalman, R. E., "Identification from real data", In: *Current Developments in the Interface: Economics, Econometrics, Mathematics*, Reidel, (Eds. M. Hazewinkel and A. H. G. Rinnooy Kan), 161—196, 1982.

Plackett, R. L. *Regression Analysis*, Oxford University Press, London, 1960.

Rao, C. R., "Some thoughts on regression and prediction", *Proc. of the Symposium to Honour Jerzy Neyman*, Warsaw, 1974.

Sclove, S. L., "(Yvs. X) oı (log Yvs. X)?", *Technometrics*, **14**, (1972).

Stigler, S. M., "Gergonne's 1815 paper on the design and analysis of polynomial regression experiments", *Historia Math.*, **1**, 431—477, (1974).

7. PARADOXES OF SUFFICIENCY

a) The history of the paradox

Sufficiency is one of the most important concepts in mathematical statistics. Its use was introduced by *R. A. Fisher* in the 1920s. He set out of the idea that, for statistical inference concerning unknown parameters, we do not always need to know all the sample elements one by one. It is enough to know some functions of the sample called sufficient statistic. E.g., in the case of a one dimensional normal distribution all the information concerning its expected value is contained in the arithmetical mean $\overline{X}$ of the sample elements $X_1, X_2, \ldots, X_n$. This follows from the fact that the distribution of $(X_1-\overline{X}, X_2-\overline{X}, \ldots, X_n-\overline{X})$ is independent of the unknown expected value and if so, no further information can be obtained about it from the random variables $X_1-\overline{X}, X_2-\overline{X}, \ldots, X_n-\overline{X}$. The mathematical definiton for sufficiency is as follows. The functions $T_1=T_1(X_1, X_2, \ldots, X_n), T_2=T_2(X_1, X_2, \ldots, X_n), \ldots, T_k=T_k(X_1, X_2, \ldots, X_n)$ are called sufficient statistic for the parameter ϑ of the common distribution of X_i if the joint distribution of $X_1, X_2, \ldots, X_n$ with given $T_1, T_2, \ldots$ $\ldots, T_k$ is independent of ϑ. Returning to the above example, the joint conditional density function of the independent variables $X_1, X_2, \ldots, X_n$ given $\overline{X}=\bar{x}$ is

$$\frac{1}{(\sqrt{2\pi}\,\sigma_0)^{n-1}\sqrt{n}}\, e^{-\frac{1}{2\sigma_0^2}\sum_{i=1}^{n}(x_i-\bar{x})^2}$$

(where σ_0 denotes the standard deviation of X_i) and this density does not depend on ϑ.

b) The paradox

It was Fisher who pointed out the following paradox of sufficiency in 1934. He studied a two-dimensional normal distribution whose coordinates were independent (for simplicity) with variance 1. Only their expected values were unknown. The arithmetical mean $\overline{X}=(\overline{X}_1, \overline{X}_2)$ of the two-dimensional sample is a sufficient statistic for the unknown pair of expected values. Suppose that the distance between the expected value (considered a vector) and the origin, i.e., $\sqrt{\vartheta_1^2+\vartheta_2^2}$ is known, say 3. Then

$$(\vartheta_1, \vartheta_2) = 3(\cos\vartheta, \sin\vartheta),$$

where ϑ is the only unknown parameter. It can be estimated by

$$\hat{\vartheta} = \operatorname{arc\,tg}\frac{\overline{X}_2}{\overline{X}_1}.$$

This is an unbiased estimator: $E(\hat{\vartheta})=\vartheta$ and its variance is $E(\hat{\vartheta}-\vartheta)^2=$ $=0.12$. It is easy to prove that the distribution of $r=\sqrt{\overline{X}_1^2+\overline{X}_2^2}$ is independent of ϑ (since the distribution of $(\overline{X}_1, \overline{X}_2)$ has a rotational symmetry around $(\vartheta_1, \vartheta_2)$) therefore, due to sufficiency, no information concerning ϑ can be gained by taking r into account. This is, however, anything but true. The expected value of $(\hat{\vartheta}-\vartheta)^2$ (i.e., the efficiency of the estimator) is strongly influenced by the knowledge of r. E.g., $E((\hat{\vartheta}-\vartheta)^2|r=1.5)=$ $=0.26$, $E((\hat{\vartheta}-\vartheta)^2|r=3)=0.12$, and $E((\hat{\vartheta}-\vartheta)^2|r=4.5)=0.08$.

c) The explanation of the paradox

Fisher's paradox points out that "having all information" can be interpreted different ways. In calculating the efficiency of estimations, the ancillary statistics (like r) may have an important role. Unfortunately, it is not always easy to decide what to take for ancillary statistic. Obviously, taking the whole sample as ancillary statistic is not worth-while. If Fisher's problem is considered from a Bayesian point of view and ϑ is supposed to be uniformly distributed on the interval $(-\pi, \pi)$ then

$$E((\hat{\vartheta}-\vartheta)^2|\overline{X}_1, \overline{X}_2) = E((\hat{\vartheta}-\vartheta)^2|r).$$

d) Remarks

The modern theory of sufficiency is due to *P. R. Halmos* and *C. L. Savage* (1949). Several interesting paradoxes were brought up in this field too. E.g., Burkholder (see the References) presented some pathological examples showing that if we add some more information to a sufficient statistic then sufficiency may get spoiled. This example totally contradicts our general view of sufficiency. In the past decade several deep papers were published in this field introducing some "regularity conditions"; these ensure the non-paradoxical (non-pathological) behaviour of sufficient statistics.

e) References

Burkholder, D. L., "Sufficiency in the undominated case", *Annals of Math. Statist.*, **32**, 1191—1200, (1961).
Burkholder, D. L., "On the order structure of the set of sufficient σ-fields", *Annals of Math. Statist.*, **33**, 596—599, (1962).
Efron, B., "Controversies in the foundations of statistics", *The American Math. Monthly*, **85**, 231—246, (1978).
Fisher, R. A., "On the mathematical foundations of theoretical statistics", *Phil. Trans. Roy. Soc. Ser. A*, **222**, 309—368, (1922).

8. PARADOXES OF THE MAXIMUM-LIKELIHOOD METHOD

a) The history of the paradoxes

One of the most efficient methods of estimating unknown parameters is the maximum-likelihood estimation. It gained ground in the twenties through the work of the English statistician *R. A. Fisher*. Though Fisher had predecessors, it was his article that made the decisive breakthrough in 1912. In order to elucidate the method, suppose, for simplicity, that the density function of the probability distribution (depending on the unknown parameter ϑ) exists and denote it by $f_\vartheta(u)$. If the sample elements $X_1, X_2, \ldots, X_n$ are independent, their joint density function is

$$\prod_{i=1}^{n} f_\vartheta(u_i).$$

Let the numbers $x_1, x_2, \ldots, x_n$ be the observed values of the sample. Then $\hat{\vartheta}$ is the maximum-likelihood estimator of ϑ, if $\hat{\vartheta}$ maximizes

$$\prod_{i=1}^{n} f_\vartheta(x_i)$$

as the function of ϑ (supposing the maximum exists and is unique). In the case of discrete random variables X_i, we maximize the joint probability $P_\vartheta\,(X_1=x_1, X_2=x_2, \ldots, X_n=x_n)$. If we estimate ϑ by the method of maximum likelihood then the probability (or probability density) that we observe $x_1, x_2, \ldots, x_n$ becomes maximal. The maximum-likelihood estimator has several good properties, this is why it is a widespread method. If, for example, $\hat{\vartheta}$ is the maximum-likelihood estimator of ϑ, then $g(\hat{\vartheta})$ is the maximum-likelihood estimator of $g(\vartheta)$. It can also be proved that under quite general conditions the maximum-likelihood estimator $\hat{\vartheta}$ behaves asymptotically as a normally distributed variable with mean ϑ and variance $\frac{1}{nI(\vartheta)}$ (see Paradox 2. Remark (*i*)); thus $\hat{\vartheta}$ is consistent and its asymptotic variance is minimal, (i.e., it is asymptotically efficient). Moreover if a sufficient estimator exists (cf. "Paradoxes of Sufficiency"), then the maximum-likelihood method gives a function of this sufficient estimator.

b) The paradoxes

(*i*) Let $X_1, X_2, \ldots, X_n$ be independent, uniformly distributed random variables on the interval $(\vartheta, 2\vartheta)$. The maximum-likelihood estimator of the unknown parameter ϑ is $\max X_i/2$. Its slight modification.

$$\hat{\vartheta} = \frac{2n+2}{2n+1} \max X_i/2$$

is an unbiased estimator of ϑ with variance $D^2(\hat{\vartheta}) = \frac{1}{4n^2}$. On the other hand, the variance of the estimator

$$\frac{n+1}{5n+4} (\min X_i + 2\max X_i)$$

is asymptotically $1/5n^2$, hence this estimator is more efficient than the maximum-likelihood estimator, whose asymptotic efficiency is maximal.

(*ii*) A very simple example can be found to illustrate that a maximum-likelihood estimator is not always consistent. Let A be the set of rational numbers between 0 and 1, and B a set of countably many irrational numbers between 0 and 1. Suppose the independent sample elements $X_1, X_2, \ldots$ $\ldots, X_n$ take only the values 0 and 1, and they take the value 1 with probability ϑ if ϑ is an element of A, and with probability $1-\vartheta$ if ϑ is an element of B. Then the maximum-likelihood estimator of ϑ is not consistent. (Though there exists a somewhat more complicated consistent estimator of ϑ.)

c) The explanation of the paradoxes

(*i*) The statistics $\xi=\min X_i$ and $\eta=\max X_i$ together contain all the information concerning ϑ, more precisely, given ξ and η the joint density function of $X_1, X_2, \ldots, X_n$ does not depend on ϑ (i.e., ξ and η together are sufficient). Thus it is natural that both the maximum-likelihood estimator and the one which turned out to be a better estimator depends only on ξ and η. Since the maximum-likelihood estimator depends only on η, which is not sufficient by itself (it does not contain all the information concerning ϑ), it is not very surprising that we could find a better estimator. This does not contradict the asymptotic efficiency of maximum-likelihood estimators since, in case of uniform distributions, the "general conditions" that assure this efficiency do not hold.

(*ii*) The explanation is quite simple: the maximum-likelihood estimator of ϑ is the relative frequency

$$\sum_{i=1}^{n} X_i/n$$

and it tends to $1-\vartheta$ if ϑ is irrational.

Though this problem is somewhat pathological, it is at least easy to understand. (The paper of *D. Basu* gives a consistent estimator of ϑ.) There exist other examples of non-consistent maximum-likelihood estimators which are less artificial, but more complicated (cf. the papers by *Neyman—Scott, Kiefer—Wolfowitz* and *Ferguson*).

d) Remarks

(*i*) There are numerous "maximum-likelihood" estimators in the statistical literature where no real maxima were found (just saddle-points) or only one of the local maxima was considered.* Though the frequent appearance of these examples is rather interesting, they cannot be considered paradoxes, only "oversights", even if they are published in first rate journals by the best mathematicians.

* One of the simplest and most important example where the local maximum is not unique is the normal distribution with unknown expectation ϑ and variance proportional to ϑ^2.

(*ii*) Examples of *J. L. Hodges* and others raised the paradox problem of superefficiency. Here we only refer to the dissertation by *L. Le Cam* and the paper by *H. Chernoff*. (An estimator of ϑ is superefficient if its distribution is asymptotically normal with mean ϑ and variance not more than the asymptotically minimal $\frac{1}{nI(\vartheta)}$ and strictly less for at least one value of ϑ.)

e) References

Bahadur, R. R., "Examples of inconsistencies of maximum likelihood estimates", *Sankhya*, **20**, 207—210, (1958).

Barnett, V. D., "Evaluation of the maximum likelihood estimator when the likelihood equation has multiple roots", *Biometrika*, **53**, 151—165, (1966).

Basu, D., "An inconsistency of the method of maximum likelihood", *Annals of Math. Statist.*, **26**, 144—145, (1955).

Berkson, J., "Minimum chi-square, not maximum likelihood!" *Annals of Statist.*, **8**, 457—487, (1980).

Boyles, R. A., Marschall, A. W., Proschan, F.; "Inconsistency of a distribution having increasing failure rate average", *Annals of Statist.*, **13**, 413—417, (1985).

Chernoff, H., "Large sample theory: parametric case", *Annals of Math. Statist.*, **27**, 1—22, (1956).

Edwards, A. V. T., "The history of likelihood", *Internat. Statist. Rev.*, **42**, 9—15, (1974).

Ferguson, T. S., "An inconsistent maximum likelihood estimate", *J. Amer. Statist. Assoc.*, **77**, 831—834, (1982).

Fisher, R. A., "On an absolute criterion for fitting frequency curves", *Messenger of Mathematics*, **41**, 155—160, (1912).

Fisher, R. A., "On mathematical foundations of theoretical statistics", *Phil. Trans. Roy, Soc. (London) Ser. A*, **222**, 309—368, (1922).

Kale, B. K., "Inadmissibility of the maximum likelihood estimation in the presence of prior information", *Canad. Math. Bull.* **13**, 391—393, (1970).

Kiefer, J., Wolfowitz, J., "Consistency of the maximum likelihood estimation in the presence of infinitely many incidental parameters", *Annals of Math. Statist.*, **27**, 887—906, (1956).

Konijn, H. S., "Note on the nonexistence of a maximum likelihood estimation", *Aust. J. Statist.*, **5**, 143—146, (1963).

Kraft, C., Le Cam, L., "A remark on the roots of the likelihood equation", *Ann. Math. Statist.*, **27**, 1174—1177, (1956).

Le Cam, L., "On some asymptotic properties of maximum likelihood estimates and related Bayes' estimates", *Univ. California Publ. Stat.*, **1**, 277—330, (1953).

Neyman, J., Scott, E. L., "Consistent estimates based on partially consistent observations", *Econometrica*, **16**, 1—32, (1948).
Norden, N. H., "A survey of maximum likelihod estimation", *Intern. Statist. Rev.*, **40**, 329—354, (1972).
Rao, C. R., "Apparent anomalies and irregularities in maximum likelihood estimation", *Sankhya*, **24**, 73—102, (1962).
Reeds, J. A., "Asymptotic number of roots of Cauchy location equations", *Annals of Statist.*, **13**, 775—784, (1985).

9. THE PARADOX OF INTERVAL ESTIMATIONS

a) The history of the paradox

The theory of interval estimation was developed basically by *R. A. Fisher* and *J. Neyman* between 1925 and 1935. Neyman's *confidence interval* contains the unknown parameter ϑ with a prescribed probability α. Let $X_1, X_2, \ldots, X_n$ denote the sequence of sample elements and let $A = A(X_1, X_2, \ldots, X_n, \alpha)$ and $B = B(X_1, X_2, X_n, \alpha)$ be such that $P(A < \vartheta < B) = \alpha$. Then (A, B) is called the α-confidence interval for ϑ. If ϑ denotes the unknown expectation of a normal distribution with standard deviation σ, then

$$P(\bar{X} - 2\sigma/\sqrt{n} < \vartheta < \bar{X} + 2\sigma/\sqrt{n}) \approx 0.95,$$

i.e.,

$$(\bar{X} - 2\sigma/\sqrt{n}, \bar{X} + 2\sigma/\sqrt{n})$$

is a 95% confidence interval for ϑ. Another type of interval estimation considers not the sample but the unknown parameter ϑ as a random variable. In this case the interval (A, B) does not depend on chance, and

$$P(A < \vartheta < B) = \alpha$$

means simply that ϑ falls into the interval (A, B) with probability α. E.g., if ϑ denotes the unknown expected value of a normal distribution then ϑ is not determined completely by the sample mean $\bar{X}$ due to random errors in measurement, this ϑ can be considered a normally distributed random variable with expected value $\bar{X}$ and standard deviation $\sigma/\sqrt{n}$. Hence

$$P(\bar{X} - 2\sigma/\sqrt{n} < \vartheta < \bar{X} + 2\sigma/\sqrt{n}) \approx 0.95.$$

This kind of interval estimates called *fiducial intervals* was introduced by *Fisher*. As we can see in case of normal distribution, confidence and fiducial intervals of expected values are formally the same only their "philosophy" is different. It was believed for a while that these two intervals were practically the same and the debate confidence contra fiducial seemed to be only theoretical. (At first it was Neyman who supported Fisher's fiducial theory the most mainly because Fisher also failed to apply Bayes' theorem.) Paradoxes of practical importance have appeared, however, rather soon. The different philosophy of Fisher and Neyman led to different results in practical application as well. In 1959 *C. Stein* pointed out an extremely paradoxical case. For simplicity, he considered confidence and fiducial intervals for which $B=\infty$ or $A=-\infty$ because these kind of intervals are determined by a single value (the other end point of the interval).

b) The paradox

Let $X_1, X_2, \dots, X_k$ be independent, normally distributed random variables with unit variance ($k \geqq 2$) and denote $\vartheta_1, \vartheta_2, \dots, \vartheta_k$ their unknown expected value. Let the distance of the vector $\vartheta=(\vartheta_1, \vartheta_2, \dots, \vartheta_k)$ from the origin be

$$|\vartheta| = \sqrt{\vartheta_1^2+\vartheta_2^2+\dots+\vartheta_k^2}.$$

Stein proved that the confidence and fiducial intervals of $|\vartheta|$ may differ extremely which results in the following paradox. Let us estimate every ϑ_i by the mean value $\bar{X}_i$ of an n-sized sample. Let the distance between the origin and the sample mean vector $(\bar{X}_1', \bar{X}_2, \dots, \bar{X}_k)$ be $|\bar{X}|= =\sqrt{\bar{X}_1^2+\bar{X}_2^2+\dots+\bar{X}_k^2}$. Then $\mathrm{P}(|\bar{X}|>|\vartheta|)>0.5$ if $\bar{X}$ is the random variable (confidence interval) whatever the value of the unknown ϑ is. On the other hand, if ϑ is the random variable (fiducial interval) then $P(|\vartheta|>|\bar{X}|)>0.5$ for any value of the sample mean $\bar{X}$. In other words, the probability that the confidence interval $(-\infty, |\bar{X}|)$ contains the unknown $|\vartheta|$ is more than 50% while it has also more than 50% chance that the random $|\vartheta|$ is contained by the (fiducial) interval $(|\bar{X}|, +\infty)$. Thus, in the confidence approach, it is favourable to bet on the inequality $|\bar{X}|>|\vartheta|$ while with the fiducial approach, it is just the other way round.

c) The explanation of the paradox

It is impossible to show all the discrepancies between the confidence and fiducial approaches in connection with Stein's problem. Here we restrict ourselves to a solution proposed by Stein himself. If the fiducial approach is applied not to the sample elements given by their coordinates but (because of the rotation symmetry of the normal distribution) to the sum of squared coordinates then fiducial intervals became equivalent to confidence intervals (see Stein's paper). Consequently, it is more advantageous to bet on "$|\bar{X}|$ is greater than $|\vartheta|$".

d) Remarks

(*i*) Let us construct an interval estimation for the unknown expected value ϑ of a normal distribution with known standard deviation σ, using the *a priori* information that ϑ is normally distributed with an expected value of μ and standard deviation of s (μ and s are known). If the mean of the n-sized sample is $\bar{X}$, then, according to Bayes' theorem, the *a posteriori* distribution of ϑ is also normal with the expected value

$$\vartheta^* = \mu + C(\bar{X} - \mu)$$

and standard deviation D, where

$$C = \frac{n/\sigma^2}{1/s^2 + n/\sigma^2} \quad \text{and} \quad D^2 = \frac{1}{1/s^2 + n/\sigma^2}.$$

Therefore $(\vartheta^* - 2D, \vartheta^* + 2D)$ is a 95% interval estimate for ϑ, because $P(\vartheta^* - 2D < \vartheta < \vartheta^* + 2D) \approx 0.95$. The lack of *a priori* information means that $s = \infty$, that is, $C = 1$. Thus $\vartheta^* = \bar{X}$ and $D^2 = \sigma^2/n$, which is just the fiducial interval. Consequently, in the case of multidimensional normal distributions, the Bayesian approach results in the same paradox as fiducial reasoning does directly. Another paradox of this type is the following (it comes from the Moscow statistical school). A machine consists of m components in series connections, thus if the i-th component works with probability p_i ($i = 1, 2, \ldots, m$) then the machine works with probability $p = p_1 \cdot p_2 \ldots p_m$. Now taking a sample of $n_1, n_2, \ldots, n_m$ elements, it turns out that all of them works perfectly. Using this information, find an interval estimate of the form

$$P(p > p^*) = \alpha.$$

Surprisingly enough, the confidence interval (i.e., the variable p^*) does not depend on m, only on

$$\min_{1 \le i \le m} n_i = n_0$$

and on the corresponding probability p_0. At the same time, in the Bayesian framework, the interval estimate for p depends on m.

(*ii*) Fisher (1890—1962) had begun to deal with interval estimates, a bit earlier than Neyman (1894—1981). Fisher even accused Neyman, who was then working in Poland, of appropriating and enlarging his ideas. At that time he had already both personal and professional conflicts with other outstanding statisticians. He hated *K. Pearson* (1857—1936) and for this reason he did not publish after 1920 in Biometrika (the leading periodical of statistics, established and edited, among others, by Pearson). Fisher transmitted his antipathy, though lessened, to K. Pearson's son E. S. Pearson (1895—1980) and his friend J. Neyman. Later, Neyman became one of the leading statisticians in the USA and their dispute turned into an Anglo—American dispute. Fisher had never liked the idea of reducing statistical conclusions to decisions with loss functions. (This "American" tendency in statistics was developed by the Hungarian *Abraham Wald* on the basis of Neumann's game theory.) The strong contrast was expressed as follows: In America (corresponding to *Peirce*'s pragmatism) it is not important what we think but what we do. In England it is just the contrary. Fisher, though his reasonings are not always convincing, is one of the greatest (if not the greatest) statisticians who has ever lived. So it is strange that he was never made professor of statistics. He did in fact become a professor at Cambridge University in 1943, but in genetics. He also became the president of the Royal Society between 1952 and 1954.

(*iii*) We are to estimate the location parameter ϑ, from the sample values $X_1, X_2, \dots, X_n$, distributed according to the exponential density function $e^{\vartheta - x}$ (if $x > \vartheta$ and 0 otherwise). The estimator

$$\hat{\vartheta} = \frac{1}{n} \sum_{i=1}^{n} (X_i - 1)$$

is unbiased, its probability density function is proportional to $(x-\vartheta+ +1)^{n-1}e^{-n(x-\vartheta+1)})$ for $x>\vartheta-1$. Using this density, we can easily determine the shortest 90% confidence interval. In the case $X_1=12$, $X_2=14$, $X_3=16$, this confidence interval is $12.1471<\vartheta<13.8264$. On the other hand, ϑ is obviously less than $X_1^*=\min X_i=12$. *Thus the shortest 90% confidence interval lies in the region where it is impossible for ϑ to be!* Jaynes emphasizes (see the reference below) that the Bayesian solution is the proper way to determine an interval estimation. If the prior density is constant, the posterior density of ϑ will be $ne^{n(\vartheta-X_1^*)}$ (if $\vartheta<X_1^*$ and 0 otherwise). The shortest posterior belt that contains $100P$ percent of the posterior probability is thus $X_1^*-q<\vartheta<X_1^*$, where $q=n^{-1}\log(1-P)$. For the above sample values $11.23<\vartheta<12.0$. From the "confidence" point of view, one can argue that $\hat{\vartheta}$ is not a sufficient statistic for ϑ while X_1^* is sufficient. The shortest confidence interval based on the sufficient statistic is the same as the Bayesian interval above. But even if we work with X_1^*, it may occur that a 90% confidence interval $(-\infty, f(X_1^*))$ lies in the negative half-line when we know (from prior information) that ϑ cannot be negative.

e) References

Bartholomew, D. J., "A comparison of some Bayesian and frequentist inference", *Biometrika*, **52**, 19—35, (1965).

Birnbaum, A., "On the foundation of statistical inference (with discussion)", *J. Amer. Statist. Assoc.* **57**, 269—326, (1962).

Box, J. F., R. *A. Fisher: The Life of a Scientist*, Wiley, New York, 1978.

Dempster, A. P., "On a paradox concerning inference about a covariance matrix", *Annals of Math. Statist.*, **34**, 1414—18, (1963).

Efron, E., "Controversies in the foundations of statistics", *The American Math. Monthly*, **85**, 231—246, (1978).

Fisher, R. A., "Inverse probabilities", *Proc. Cambridge Phil. Soc.*, **26**, 528—535, (1930).

Fisher, R. A., *Statistical Methods and Scientific* Inference, Oliver and Boyd, London, 1956.

Jaynes, E. T., "Confidence intervals vs Bayesian intervals", In*: Foundations of Probability Theory, Statistical Inference, and Statistical Theories of Sciences*, eds. by Harper, W. L. and Hooker, C. A.; D. Reidel Publishing Co., Dordrecht, Holland, 1976.

Jeffreys, H., *Theory of Probability*, Clarendon Press, Oxford, 1967.

Kendall, M. G., "R. A. Fisher 1890—1962", *Biometrika*, **50**, 1—16, (1963).
Neyman, J. "On the two different aspects of representative method: The method of stratified sampling and the method of purposive selection", *Estadística*, **17**, 587—651, (1959).
Robinson, G. K., "Some counterexamples to the theory of confidence intervals", *Biometrika*, **62**, 155—162, (1975).
Stein, C., "An example of wide discrepancy between fiducial and confidence intervals", *Annals of Math. Statist*, **30**, 877—880, (1959).
Stone, M., "Strong inconsistency from uniform priors", *J. Amer. Statist. Assoc.*, **71**, 114—116, (1970).

10. THE PARADOX OF TESTING A HYPOTHESIS

a) The history of the paradox

It is very difficult to say anything definite about the first attempts to test a statistical hypothesis but in his monograph *B. V. Gnedenko* states that in ancient China as early as 2238 BC censuses showed that the birth rate of boys was 50%. *John Arbuthnot* (1667—1735), an English matematician, doctor and writer, was the first to point out (in 1710) that the hypothesis of equality of the birth rate of boys and girls must be rejected, since, according to the demographic data over an 82 year period (available at that time), more boys than girls were born each year. If the probability that a newborn baby is a boy was 1/2, the experience of 82 years would be so improbable ($1/2^{82}$) that it can be considered almost impossible. So Arbuthnot was the first who rejected a natural statistical hypothesis. This (not mathematical) paradox aroused the interest of *Laplace*. In 1784 he was surprised to find that the birth rate of boys was approximately equal 22/43 in several different places, whereas the same ratio was 25/49 in Paris. Laplace was intrigued by such a remarkable difference, but he shortly found a rational explanation: the total number of births in Paris included all foundlings and the surrounding population had a preference for abandoning infants of one sex. When Laplace eliminated the foundlings from the total number of births, the birth rate of boys came close to the number 22/43.

In 1734, *D. Bernoulli* won a prize from the French Academy for an essay on the orbits of planets. Constructing a hypothesis test, Bernoulli attempted to show that the similarity of planetary orbital planes would have been most unlikely to have occurred by chance. Using the right hand rule, each orbit corresponds to a point on a unit sphere, and he tested the hypothesis that these points were drawn from a uniform distribution on the unit sphere. In 1812, Laplace analyzed a similar problem. He attempted to apply statistical methods to decide which hypothesis should be accepted: that the comets are regular members of the solar system or that they are only "intruders". In the latter case the angles between the orbital planes of comets and the ecliptic would be uniformly distributed between 0 and $\pi/2$, and this was exactly the mathematical form of Laplace's hypothesis. (He found that comets are *not* regular members of the solar system.) The modern theory of testing statistical hypotheses was initiated by *K. Pearson, E. S. Pearson, R. A. Fisher* and *J. Neymann.*

Suppose we have to test the hypothesis that the probability distribution of a random variable is F. (In the problem of Laplace, F was the uniform distribution on the interval $[0, \pi/2]$.) For this "goodness of fit" problem *K. Pearson, H. Cramér, R. von Mises, A. N. Kolmogorov, N. V. Smirnov* and others who followed them worked out several different tests, and it became necessary to compare their efficiency. E. S. Pearson and J. Neyman made the first move to solve the theoretical and practical problem of finding the best decision methods. First they introduced the notion of *alternative hypothesis,* which is not necessarily the opposite of the original, *null hypothesis*. For example, consider a random variable which is normally distributed with unit variance and unknown expectation; if the *null hypothesis* is that "the expectation is -1" and the alternative hypothesis is that "the expectation is $+1$", then the two hypotheses obviously do not cover all the possibilities. In connection with these simple hypotheses (where both the null hypothesis and the alternative hypothesis contained a single distribution) Neyman and Pearson showed in 1933 that there exists a most powerful test in the following sense. When a statistical test is performed, two kinds of errors are possible. We may reject the null hypothesis when it is true, making a *type I error* (or error of the first kind). On the other hand, we may accept the null hypothesis when it is false, making a *type II error* (or error of the second kind). A decision method (test) based on a sample of given size is called most powerful if, for an arbitrary fixed probability of a type I error, the probability of a type II error is as small as possible. (If the size of the sample is given, the sum of the probabilities of the two types of errors cannot be made arbitrarily small. This fact is a kind of *uncertainty principle* in hypothesis testing.) Suppose, for simplicity, that both distributions (in the null hypothesis and in the alternative hypothesis) have density functions. Then, according to the fundamental principle of Neyman and Pearson, there is a most powerful test of the following form. Denote by f_0 and f_1 the density function of the sample $X=(X_1, X_2, \ldots, X_n)$ under the null hypothesis and the alternative hypothesis, respectively. We accept the null hypothesis if and only if

$$\frac{f_1(X)}{f_0(X)} < c, \text{ where } c \text{ is a suitable constant.}$$

(For simplicity we suppose that the probability of $f_1(X)/f_0(X)=c$ is 0.) The theory of Neyman and Pearson became fundamental in testing hypotheses, but not without paradoxes. Herbert Robbins showed in 1950 that there is a test which is in a sense more powerful than the most powerful test of Neyman and Pearson.

b) The paradox

Suppose X is a normally distributed random variable with expectation ϑ and variance 1. Let the null hypothesis be $\vartheta=-1$ and the alternative hypothesis $\vartheta=+1$. On the basis of a single sample element X, the most powerful test of the null hypothesis against the alternative hypothesis is the following: we accept it and reject the alternative hypothesis if $X\leqq 0$, otherwise we reject it and accept the alternative hypothesis. In this case the probability of both kinds of errors is approximately 16%, since

$$P(X>0|\vartheta=-1) = P(X<0|\vartheta=+1) =$$

$$= \frac{1}{\sqrt{2\pi}} \int_0^\infty e^{-(x+1)^2/2}\,dx = 0.1587\ldots \approx 0.16.$$

If we apply this test in N independent cases, then for large N the expected number of false decisions is approximately $0.16N$. Since we have used the most powerful test in each case, one might think that the average number of false decisions cannot be smaller than $0.16N$. The following method of Robbins shows that, paradoxically, this is not the case.

Let $\bar{X}$ be the average of the observations $X_1, X_2, \dots, X_N$. Robbins' test is the following:

$$\text{if} \quad \bar{X} < -1, \quad \text{then} \quad \vartheta_i = -1 \quad \text{for all} \quad i = 1, 2, \dots, N,$$

$$\text{if} \quad \bar{X} > +1, \quad \text{then} \quad \vartheta_i = +1 \quad \text{for all} \quad i = 1, 2, \dots, N,$$

and finally

$$\text{if} \quad -1 \leqq \bar{X} \leqq +1, \quad \text{then} \quad \vartheta_i = -1 \quad \text{or} \quad \vartheta_i = +1,$$

depending on whether the inequality

$$X_i \leqq \frac{1}{2} \ln \frac{1-\bar{X}}{1+\bar{X}}$$

holds or not. This method is very surprising because it connects independent problems. For large N, (e.g., for $N=100$), if the true ratio of $\vartheta_i = +1$ to $\vartheta_i = -1$ is 0 or 1, then Robbins' procedure decides with 100% certainty; for a ratio 0.1 or 0.9 the probability of error (of both types) is 7%; for a ratio 0.2 or 0.8 the probability of false decisions is 11%; for a ratio 0.3 or 0.7 it is 14% and even for the ratio 0.4 or 0.6 the percentage of errors is still smaller than the level 16% of the most powerful test. Robbins' method becomes less efficient than the most powerful test only in the case of ratio near to 0.5.

c) The explanation of the paradox

Robbins' paradox shows that even when we have to make decisions about accepting or rejecting products from different factories working independently, the total number of false decisions will be fewer on average, if we do not make our decisions independently of each other. Since this is essentially the same problem as Stein's paradox on admissible estimators of the expectation, here we only refer to its explanation in Section 2 and to Robbins' fundamental paper.

d) Remark

Other paradoxes on hypothesis testing will be discussed in Sections 12 and 13 among the quickies.

e) References

Lehmann, E. L., *Testing Statistical Hypotheses*, Wiley, New York, 1959.
Neyman, J., "Two breakthrough in the theory of statistical decision making", *Internat. Statist. Rev.*, **30**, 11—27, (1962).
Neyman, J., "Egon S. Pearson (August 11, 1895—June 12, 1980)", *Annals of Statist.*, **9**, 1—2, (1981).
Neyman, J., Pearson, E. S., "On the problem of the most efficient test of statistical hypotheses", *Phil. Trans. Roy. Soc.*, **231**, 289—337, (1933).
Robbins, H., "Asymptotically subminimax solution of the compound statistical decision problem", *Proc. 2nd Berkeley Symp. on Math. Statist. and Prob.*, 131—148, Univ. Calif. Press, Berkeley, (1950).

11. RÉNYI'S PARADOX OF INFORMATION THEORY

a) The history of the paradox

One of the main tasks of information theory is to measure the amount of information. The pioneers of this discipline of mathematics (*C. Shannon, N. Wiener* and others) realized that the amount of information is measurable by a scalar independent of the actual meaning and form of information, like the volume of liquid is independent of its shape. The unit of information is the information content of an answer "yes" or "no". In binary code this information can be given by a single digit (e.g., 1 for "yes", 0 for "no"), which is called a *bit* (abbr. of binary digit). What makes this abbreviation especially suitable is its meaning as a normal word. The content of information is measured by the average amount of binary numbers needed to express the information. If a random variable can take only a finite or countably infinite number of values with positive probabilities $p_1, p_2, \ldots$ then, according to *Shannon's formula*, its information content is

$$H = H(p_1, p_2, \ldots) = \sum_i p_i \log p_i^{-1} \qquad \text{(bit)}$$

where (in this passage) log stands for logarithms with base 2. H is called the *entropy* of the probability distribution $p_1, p_2, p_3, \ldots$. This is the average length of the most economical code combinations by means of which the outcomes of events with probabilities $p_1, p_2, \ldots$ can be described. Another important notion of information theory is the information gain. If the observation of a random variable (or event) changes the probability distribution $p_1, p_2, p_3, \ldots$ to $q_1, q_2, q_3, \ldots$ then the amount of the information gained is

$$\sum_i q_i \log \frac{q_i}{p_i}.$$

Now let the unknown parameter ϑ of a probability distribution be a random variable (according to the Bayesian view of mathematical statistics). For simplicity, suppose that ϑ can take only a finite or countably infinite number of values with probabilities $p_1, p_2, p_3, \ldots$. Thus the entropy of ϑ is $H(\vartheta)=H(p_1, p_2, p_3 \ldots)$. Suppose moreover that the random sample $X=(X_1, X_2, \ldots, X_n)$ can also take only finite or countably infinite number of different values with positive probabilities $q_1, q_2, q_3, \ldots$. Finally, let r_{jk} denote the probability that ϑ takes the jth value (whose probability is p_j) and at the same time X the kth value (whose probability is q_k). Then the amount of information concerning ϑ and obtained by observing X is

$$I(X, \vartheta) = \sum_{j,k} r_{jk} \log \frac{r_{jk}}{p_j q_k}.$$

A function $f(X)=f(X_1, X_2, \ldots, X_n)$ of the sample X is called sufficient if $I(f(X), \vartheta)=I(X, \vartheta)$, i.e., if $f(X)$ contains as much information concerning ϑ as the original sample X does. If f is not necessarily sufficient, the ratio $I(f(X), \vartheta)/I(X, \vartheta)$ gives the proportion of information concerning ϑ that can be obtained from the sample if $f(X)$ is used instead of the complete sample. The property that, by taking more and more observations, we can obtain at last all the information concerning ϑ can be expressed in the language of information theory as follows. If the observations $X_1, X_2, \ldots$ are independent, identically distributed random variables whose distribution F_ϑ is different for different values of the parameter ϑ, then

$$\lim_{n\to\infty} I((X_1, X_2, \ldots, X_n), \vartheta) = H(\vartheta).$$

A. Rényi's paradox discussed below comes from the application of information theory in testing hypotheses.

b) The paradox

By observing the random variable X which is in connection with the event A, we would like to guess whether A has occurred or not. If the probability of A is $P(A)=p$ then the content of information of the event A is $H(p, 1-p)$. Having observed the variable X, the amount of information still missing is $\bar{H}_X = E(H(P(A|X), 1-P(A|X)))$, where $P(A|X)$ stands for the conditional probability of the event A given X. Consequently, the content of information concerning A if X was observed is

$$I(X, Y) = H(p, 1-p) - H_X.$$

Observing X, let $d(X)=1$ if we decide that A has occurred and $d(X)=0$ in case of the complement of A, i.e., $\bar{A}$. The probability of a wrong decision (error) is

$$\delta = pP(d(X) = 0|A) + (1-p)P(d(X) = 1|\bar{A}).$$

It is easy to prove (e.g., by the fundamental result of Neyman and Pearson; see II. 10.) that no decision can have less error δ than the following "standard decision":

$$d_0(X) = \begin{cases} 1 & \text{if} \quad P(A|X) > P(\bar{A}|X), \\ 0 & \text{if} \quad P(A|X) < P(\bar{A}|X). \end{cases}$$

If $P(A|X) = P(\bar{A}|X)$ then let $d_0(X)=1$ with probability p and 0 with probability $1-p$. The paradox appearing here is the following. Let $Y=d_0(X)$. In this case the information content of Y concerning A is:

$$I(Y, A) = H(p, 1-p) - \bar{H}_Y.$$

Y is a function of X, therefore $I(Y, A) \leqq I(X, A)$. The equality holds if and only if $P(A|X)$ can take only two different values, i.e., generally, X

contains more information concerning A than Y. Still, knowing X, we cannot make a better decision concerning A than if we know only $Y=$ $=d_0(X)$. From this it follows that while X contains generally more information on A than Y, it is impossible to utilize this extra information.

c) The explanation of the paradox

The extra information can be utilized by observing another random variable. Let, e.g., $Z=X+U$, where U is the indicator variable of the event A. This means that $U=1$ if A occurs and zero otherwise. Obviously, by observing X and Z simultaneously, we get full information concerning A that is the latent extra information in the value of X concerning A can be made free by observing the auxiliary variable Z.

d) Remarks

Information theory is in close connection with several practical problems, e.g., with the optimal methods of telecommunication or the foundations of biology (see the references below).

e) References

Ash, B. B., *Information Theory,* Wiley, New York, 1966.

Brioullin, L., *Science and Information Theory,* New York, Academic Press, 1956.

Kullback, S., *Information Theory and Statistics,* Wiley, New York, 1959.

Quaster, H. (ed.), *Information Theory in Biology,* Univ. Illinois Press, Urbana, 1953.

Rényi, A., *Notes on Information Theory,* Akadémiai Kiadó, Budapest, 1984.

Shannon, C., Weaver, W., *The Mathematical Theory of Communication,* Univ. Illinois Press, Urbana, 1949.

Turán, P. (ed.), *Selected Papers of Alfréd Rényi,* Akadémiai Kiadó, Budapest, Vol. III., 442, 1976.

Yaglom, A. M., Yaglom, I. M., Hinčin, A. Ya., *Mathematical Foundation of Information Theory,* (in Hungarian) Műszaki Könyvkiadó, Budapest, 1959.

12. THE PARADOX OF STUDENT'S t-TEST

a) The history of the paradox

In the classical theory of mathematical statistics the sample elements (observations) were considered to be given in advance. One of the most important branches of modern statistics is based on the recognition that the sample size should not be fixed in advance; instead it should depend on the result of earlier observations. Thus the sample size also depends on chance. This idea of sequential sampling evolved gradually from the results of *H. F. Dodge* and *H. G. Romig* (1929), *P. C. Mahalanobis* (1940), *H. Hotelling* (1941) and *W. Bertky* (1943), but the real founder of the sequential theory of mathematical statistics was *A. Wald* (1902—1950). His sequential likelihood ratio test (1943) was a decisive discovery which enabled (in typical cases) a 50% saving on the average number of observations (with the same probabilities of errors). No wonder, Wald's discovery was classified as "restricted" during World War II. His fundamental book "Sequential analysis" was published only in 1947. A year later Wald and *J. Wolfowitz* proved that no other method can save more sample elements than the sequential likelihood ratio test. Paradoxes found their way into this field, too. Here we shall discuss the paradox noted by *C. Stein*, though it refers only to a two-stage decision, not a sequential one.

b) The paradox

Let $X_1, X_2, \dots, X_n$ be a sample of independent, normally distributed random variables with the same unknown expectation ϑ and the same unknown standard deviation σ. On the basis of this sample, we want to decide between the following null hypothesis and alternative hypothesis. The null hypothesis states that $\vartheta=\vartheta_0$ (where ϑ_0 is a given number), while the alternative hypothesis states that $\vartheta\neq\vartheta_0$. Let

$$\overline{X} = \frac{1}{n}\sum_{i=1}^{n} X_i, \quad D^{*2} = \frac{1}{n-1}\sum_{i=1}^{n}(X_i-\overline{X})^2$$

and

$$t_n = \frac{\overline{X}-\vartheta_0}{D^*/\sqrt{n}}.$$

The usual way of making a decision between the two hypotheses is the Student's t-test. According to the t-test, the null hypothesis should be accepted or rejected depending on whether the value of t_n is near enough to 0 or not. *G. B. Dantzig* showed in 1940 that given the probability of the type I error, the probability of the type II error depends on the unknown standard deviation σ for any decision method. Paradoxically, five years later C. Stein proved that if the sample size n was not fixed in advance but depends on the sample elements which have already been chosen (as in Wald's sequential analysis), then there does exist a t-test, where (given the probability of the type I error) the probability of the type II error does not depend on the unknown standard deviation σ (it depends only on the difference $\vartheta - \vartheta_0$).

c) The explanation of the paradox

In the first step choose a sample $X_1, X_2, \ldots, X_{n_0}$, where n_0 is a fixed number. The empirical sample variance is

$$s^2 = \frac{1}{n_0 - 1}\left\{\sum_{i=1}^{n_0} X_i^2 - \frac{1}{n_0}\left(\sum_{i=1}^{n_0} X_i\right)^2\right\}.$$

Let the size n of the entire sample depend on the magnitude of s and on a previously fixed number z, in the following way:

$$n = \max\left\{\left[\frac{s^2}{z}\right] + 1,\ n_0 + 1\right\},$$

where the brackets [] denote the integer part of a real number. Choose the positive numbers $a_1, a_2, \ldots, a_n$ such that

$$\sum_{i=1}^{n} a_i = 1, \quad a_1 = a_2 = \ldots = a_{n_0}$$

and

$$s^2 \sum_{i=1}^{n} a_i^2 = z,$$

and decide between the hypotheses on the basis of the following statistics:

$$t^* = \frac{\sum_{i=1}^{n} a_i X_i - \vartheta_0}{\sqrt{z}} = t + \frac{\vartheta - \vartheta_0}{\sqrt{z}},$$

where

$$t = \frac{\sum_{i=1}^{n} a_i (X_i - \vartheta)}{\sqrt{z}}.$$

Obviously, if s is given, the random variable t is normally distributed with expectation 0 and variance

$$\sigma^2 \sum_{i=1}^{n} a_i^2/z = \sigma^2/s^2.$$

On the other hand, the distribution of $(n_0 - 1)s^2/\sigma^2$ (for arbitrary σ) is the same as the distribution of the sum of squares of $n_0 - 1$ independent standard normal random variables (the $\chi^2_{n_0-1}$ chi-square distribution) which is independent of σ. Therefore the distribution of t is also independent of σ, so t^* depends only on $\vartheta - \vartheta_0$ and not on σ.

d) Remarks

(*i*) The random variable t_n is not normally distributed because D^* is not a number, but a random variable. (If the value of the standard deviation were known and we substituted this value for D^*, then t_n would have a standard normal distribution.) This remarkable observation and the analysis of the random variable t_n was published in 1908 by *Student*, alias *William D. Gosset*. (He worked for the Guinness brewery in Dublin from 1899 and his boss insisted that Gosset should write under a pseudonym.) For a long time nobody recognized the importance of Student's paper. (According to Student, even as late as 1922 *R. A. Fisher* was the only person who ever applied the t distribution; in fact it was Fisher who denoted Student's distribution by t for the first in his book published in 1925. Student himself used the letter z to denote, not exactly t_n, but $(n-1)t_n$.)

(*ii*) The optimal stopping of sampling sequences in sequential analysis was the root of the modern theory of optimal stopping of different processes. If we consider sampling as a process, we connect mathematical statistics and the theory of stochastic processes, which will be discussed in the following chapter. This connection proved an advantage for both areas. Nowadays Wald's fundamental theorems in sequential analysis are special cases of the general theory of stopped stochastic processes (see the book by Shirjaev).

e) References

Chow, Y., Robbins, H., Siegmund, D., *Great Expectations: Theory of Optimal Stopping*, Houghton Mifflin Co., Boston, 1971.
Fisher, R. A., *Statistical Methods of Research Workers*, Oliver and Boyd, Edinburgh, 1925.
Shirjaev, A. H., *Statistical Sequential Analysis*, Nauka, Moscow, 1976 (in Russian).
Stein, C., "A two sample test for a linear hypothesis whose power is independent of the variance", *Annals of Math. Statist.*, **16**, 243—258, (1945).
Student, "The probable error of a mean", *Biometrika*, **6**, 1—24, (1908).
Wald, A., "Sequential analysis of statistical data: Theory", *Restricted Report*, Sept., 1943.
Wald, A., *Sequential Analysis*, Wiley, New York, 1947.

13. QUICKIES

a) The paradox of the typical and average

The notion of average, e.g., the average salary is often used as a synonym of typical. As a matter of fact, if there are only a few extremely rich and a great many poor families in a certain country having correspondingly enormous or small incomes then the arithmetical means of their incomes is not at all typical. The median of incomes, e.g., gives a much more realistic picture. (The median means that just the same number of people have incomes larger than the median as smaller.) Besides average salary there are other misleading averages. One of these is the "average man" *(l'homme moyen)*. It is no wonder that the Belgian *L. A. J. Quételet*'s study on this subject became the source of stormy debates. The worst about the "average man" is not his greyness but the discrepancies that arise. E.g., the average height does not correspond to the average weight, etc. For this reason alone we have to doubt the truth of the words of J. Reynolds (the first president of the Royal College of Fine Arts) when he said that the source of beauty is the average.

(Ref.: Quételet, L. A. J., *Essai de Physique sociale*, (1835); *L'homme moyen*, Physique Sociale, Vol. 2, Bruxelles, 1869.)

In spite of its inconsistencies, Quételet's book of 1835 is considered a milestone if not the starting point in the quantitative analysis of human social properties. *F. Galton, K. Pearson* and *F. Edgeworth* all appreciated Quételet as the genius pioneer of regression type thoughts. It was due to his book that Galton began his scientific research. Quételet, however, had other scientific merits, too. In 1820 he founded the Royal Belgian Observatory and became its first director. He was an excellent organizer too: the Statistical Society in London was set up at his suggestion in 1834, and it was also he who suggested that the first International Congress of Statistics should be convened in Brussels in 1853.)

b) The paradox of estimation

The square of an estimate is generally not the same as the estimate of the square. If, e.g., a parameter is estimated by $\overline{X}$, that is, the mean of the observed values $X_1, X_2, \ldots, X_n$ then the obvious estimate of the square parameter is $\overline{X}^2$, which generally differs from the mean of the square of the observed values. The same is true if the square is replaced by any nonlinear function.

(Ref.: Carnap, R., *Logical Foundation of Probability*, Routledge and Kegan Paul Ltd. Broadway House, London, 1950.)

c) The paradox of accurate measurement

Our task is, e.g., to determine the length of two different rods by two measurements. The instrument we may use measures length with random error whose standard deviation is σ. Paradoxically, the best method is not measuring the rods one by one. The standard deviation of the result will be less if firstly the total length (T) is measured by putting the rods end to end and then side by side and so measuring the difference of their length (D). The approximate length of the rods is

$$\frac{T+D}{2} \quad \text{and} \quad \frac{T-D}{2}, \quad \text{respectively.}$$

The standard deviation of these lengths is $\sigma/\sqrt{2}$ which is really less than σ.

(Ref.: Hotelling, H., "Some improvements in weighing and other experimental techniques", *Annals of Math. Statist.*, **15**, 297—306, (1944).)

d) The paradoxical estimation of probability

The usual estimation of an unknown probability is the relative frequency. For example, if we toss a coin a hundred times and obtain tails 47 times then the probability of tossing tails is estimated at 47/100. However, if we toss a more or less fair coin 10 times but do not obtain any tails, it is unreasonable to consider the probability of tails to be 0. If we have some *a priori* information (e.g., the coin is more or less fair) then estimating by the relative frequency is generally not the best method. Our *a priori* information can be well expressed by the beta distribution depending on two parameters a and b. (The density function of the beta distributions is 0 outside the interval (0, 1) and proportional to $x^{a-1}(1-x)^{b-1}$ on (0, 1); ($a>0$, $b>0$.) The expected value and the variance of the beta distribution is

$$m=\frac{a}{a+b} \quad \text{and} \quad d=\frac{ab}{(a+b)^2(a+b+1)}, \quad \text{respectively.}$$

Thus solving this system of equations, our *a priori* information concerning m and d can be expressed by a and b (e.g., if the coin is fair then $m=1/2$, thus $a=b$). If the *a priori* distribution is beta with parameters (a, b) then, by Bayes' theorem, the *a posteriori* distribution will be of beta type, too. (This property makes the beta distribution widely applicable.) If an event with unknown probability occurs k times out of n experiments then the parameter of the *a posteriori* beta distribution will be $(a+k, b+n-k)$, thus the *a posteriori* expected value is

$$M = \frac{a+k}{a+b+n},$$

which contains more information and is a better estimation of the unknown probability than the relative frequency k/n. Of course if n is large enough then M hardly differs from the relative frequency, but, e.g., in case $n=10$, $k=0$ and $a=b=100$, we get

$$M = \frac{100}{210} \approx 0.48,$$

whereas the relative frequency is 0, which is absolute nonsense.

(Ref.: Good, I. J., *The Estimation of Probability*, MIT Press, Cambridge, 1965.)

e) The more the data the worse the conclusions

Quite obviously, more data enables us to calculate better results. The following paradox, however, seems to show just the contrary. Let X_1, X_2, and X_3 denote independent random variables and suppose that the distributions of X_1 and X_2 are the same: both X_1 and X_2 are equal either to 0 or to 2 with the same probability, hence both have the same expected value, namely 1. Let further X_3 be equal either to 1 or to 2.5 with equal probability, so its expected value is 1.75. All this information is unknown to a mathematician who takes samples from these distributions in order to select the one with the greatest expected value. The most obvious choice is the distribution whose sample mean is the greatest. Take first a sample of a single element from every distribution. The probability of the correct selection is then

$$P(X_1 < X_3 \quad \text{and} \quad X_2 < X_3) =$$

$$= P(X_3 = 2.5) + P(X_3 = 1)\, P(X_1 = 0).\ P(X_2 = 0) = \frac{5}{8}.$$

Now what happens if one of the sample sizes (e.g., that of X_3) is increased to 2 (the others remain unchanged)? The probability of the correct selection is then:

$$P(X_1 < \overline{X}_3 \quad \text{and} \quad X_2 < \overline{X}_3) = \frac{7}{16}$$

where $\overline{X}_3$ is the arithmetical mean of the two elements of the sample. Thus the probability of the correct selection has decreased from more than 50% to less than 50%.

(Ref.: Chius, W. K., Lam, K., *The American Statistician,* 1975.)

F. Y. Edgeworth's famous paradox (1883) concerns a similar problem: if X_1 and X_2 are independent random variables with the same density function $f(x-\vartheta)$ symmetrical to ϑ then it could easily happen that X_1 is closer to ϑ than $\overline{X}=(X_1+X_2)/2$ in the sense that

$$P(|X_1-\vartheta| \leqq \varepsilon) > P(|\overline{X}-\vartheta| \leqq \varepsilon)$$

for any positive ε. This is the case, e.g., when

$$f(x) = \frac{3}{2}\frac{1}{(1+|x|)^4},$$

because the density function of X_1 is greater, at the point ϑ, than that of $\overline{X}$.

(Ref.: Stigler, S. M., "An Edgeworth curiosum", *Annals of Statist.* **8**, 931—934, (1980).)

f) The paradox of equality of expected values

Let the expected values of three normal random variables with the same variance be m_1, m_2 and m_3. It can happen that, applying, e.g., Student's t-test, we accept the hypotheses $m_1=m_2$ and $m_2=m_3$ (at a certain confidence level) but reject $m_1=m_3$! (The problem of equality of expected values is tested on the basis of two n-sized samples by the statistic

$$t = \frac{\overline{X}-\overline{Y}}{2D^*},$$

where $\overline{X}$ and $\overline{Y}$ are the sample means and D^* is the empirical standard deviation. The statistic t has a Student's distribution with the parameter $2n-2$.) This paradox was the starting point of many researches on simultaneous testing (analysis of variance, etc.).

(Ref.: Hodges, J. L., Lehmann, E. L., "The efficiency of some nonparametric competitors of the *t*-est", *Annals of Math. Statist.*, **27**, 324—335, (1956).
Scheffé, H., *The Analysis of Variance*, Wiley, New York, 1959.)

g) A paradoxical estimation for the expectation of a normal distribution

We wish to estimate the unknown expected value of a (one-dimensional) normal distribution with unit standard deviation from an n-sized sample. It is known that the arithmetical mean $\overline{X}$ of the sample is an estimator of many favourable properties. It is, e.g., unbiased, has minimal variance, admissible, and minimax under the quadratic loss function. In spite of these properties, if our aim is only to give as close estimator for ϑ as possible with given probability then there exists a better estimator $\hat{\vartheta}$, i.e.

$$P(|\hat{\vartheta}-\vartheta| < |\overline{X}-\vartheta|) > \frac{1}{2}$$

for any possible ϑ. This type of estimator is the following

$$\hat{\vartheta} = \overline{X}-\frac{1}{2\sqrt{n}}\min(\sqrt{n}\,\overline{X},\, \Phi(-\sqrt{n}\,\overline{X}))$$

if $\overline{X}\geqq 0$ and

$$\hat{\vartheta} = \overline{X}+\frac{1}{2\sqrt{n}}\min(\sqrt{n}\,\overline{X},\, \Phi(-\sqrt{n}\,\overline{X}))$$

if $\overline{X}\leqq 0$, where Φ denotes the distribution function of the standard normal distribution.

h) A paradox on testing normality

We want to test whether a given sample $X_1, X_2, \ldots, X_n$ may or may not come from a distribution with continuous distributon function $F(X)$. Let

$$F_n(x) = \frac{1}{n} \sum_{i:X_i<x} 1$$

denote the empirical (cumulative) distribution function of the sample. According to Kolmogorov's theorem, if the hypothesis is true then

$$\lim_{n\to\infty} P\left(\sqrt{n} \sup |F_n(x) - F(x)| < z\right) =$$

$$= \sum_{j=-\infty}^{+\infty} (-1)^j e^{-2j^2z^2} = K(z).$$

Using this theorem, it is easy to construct a test with confidence level α. (If $\sqrt{n} \sup |F_n(x) - F(x)|$ exceeds a critical value z_0 for which $K(z_0) = \alpha$ then the hypothesis will be rejected.) If the normality of a probability distribution is to be tested, then first the expected value and the standard deviation of the hypothetical normal distribution should be estimated from the sample by the usual $\overline{X}$ and D^*. Secondly, the above Kolmogorov test should be applied for a normal distribution $F(x)$ with the expected value $\overline{X}$ and the standard deviation D^*. Then we might think that if n is large enough the substitution of the unknown parameters by $\overline{X}$ and D^* does not cause any essential difference. The difference is, however, significant. E.g., at a 95% confidence level the critical value z_0 in Kolmogorov's test is 1.36, while a precise analysis would show that the correct critical value is 0.9. The explanation of this paradox is rather simple. Due to substitutions, $F(x)$ and the empirical $F_n(x)$ have come closer to each other, so it is advisable to choose a smaller critical value.

(Ref.: Durbin, S., "Some methods of constructing exact tests", *Biometrika*, **48**, 41—45, (1961).

Kac, M., Kiefer, J., Wofrowitz, J., "On tests of normality and other tests of goodness of fit based on distance methods", *Annals of Math. Statist.*, **26**, 189—211, (1955).

Sarkadi, K., "On testing for normality", *Proc. 5th Berkeley Symp. on Math. Statist. and Prob.*, I. 373—387, 1967.)

i) A paradox of linear regression

Suppose that a random quantity X can only be measured with an error ε having an expected value of 0. In other words, the result of the measurement $Y=X+\varepsilon$ is a very simple linear regression on X. Is there a "better estimator" for X than the measured Y? Surprisingly, in some special cases the answer is affirmative. At least, there exists an estimator $\hat{X}$ for which $E(\hat{X}-X)^2$ is less than $E(Y-X)^2=E\varepsilon^2$. Suppose, e.g., that X and ε are uncorrelated and the regression function of X on Y is also linear. Then

$$\hat{X} = \frac{D^2(\varepsilon)}{D^2(Y)} E(X) + \frac{D^2(X)}{D^2(Y)} Y$$

is a better estimator than Y. (In the extreme case of $D^2(\varepsilon)=0$ we get $\hat{X}=Y$.)

j) Sethuraman's paradox

There exists statistical functions A and B such that the unbiased estimator of the unknown parameter ϑ, based on A have smaller variance than the estimator based on B (whatever the true value of ϑ); on the other hand, when testing the null hypothesis $\vartheta=\vartheta_0$ (e.g., against the alternative hypothesis $\vartheta>\vartheta_0$), a test based on the function A is not necessarily better than a test based on B; the latter can be better locally (in a neighbourhood of the null hypothesis). If, for example, the sample elements $X_1, X_2, \ldots, X_n$ are uniformly distributed on the interval $(\vartheta; 2\vartheta)$, the maximum likelihood estimator of ϑ is

$$U = \frac{1}{2} \max(X_1, X_2, \ldots, X_n),$$

and a slight modification leads to the unbiased estimator

$$B = \frac{2n+2}{2n+1} U.$$

The following estimator A is also unbiased but with smaller variance.

$$A = \frac{n+1}{5n+4}(4U+V), \quad \text{where} \quad V = \min(X_1, X_2, \ldots, X_n).$$

To test the hypothesis $\vartheta=\vartheta_0$, however, the method based on B is locally more powerful.

(Ref.: Sethuraman, J., "Conflicting criteria of 'goodness' of statistics", *Sankhya*, **23**, 187—190, (1961).)

k) A paradox on minimax estimation

The notion of minimax estimation was introduced in Remark (*ii*) of the Paradox I. 12. Minimax estimations usually suit common sense. The following example of *H. Rubin,* however, shows the contrary. The only minimax estimator of the unknown probability $p\neq 0$ is the identically 0 estimator if the loss function is $L(p, c)=\min\left((p-c)^2/p^2; 2\right)$. So, no matter what the sample was it is reasonable to estimate the unknown parameter by 0 (a value which was ruled out in advance among the possible values of p).

Remark: If the loss function is $L(p, c)=(p-c)^2$, the minimax estimator is

$$p^* = \frac{\varkappa+\sqrt{\frac{n}{2}}}{n+\sqrt{n}},$$

where n is the sample size and $\varkappa$ is the frequency of the event with unknown probability.

l) Robbins' paradox

It is well-known that the "best" estimator of the parameter of a Poisson distribution on the basis of a single observation X is just X. (This is a minimum variance unbiased, maximum likelihood estimator.) But how can we estimate the parameters $\vartheta_1, \vartheta_2, \ldots, \vartheta_k$ of k independent Poisson distributions on the basis of the corresponding observations $X_1, X_2, \ldots$ $\ldots, X_k$ if we want

$$E\left(\sum_{i=1}^{k}(\hat{\vartheta}_i-\vartheta_i)^2\right)$$

to be minimal? Is there a better estimator than $\hat{\vartheta}_i = X_i$? It was *H. Robbins* who first pointed out that, though the k Poisson distributions are independent, it is still possible to find better estimators which take into account not only the observations of "their own" (i.e., corresponding), but also of the others. According to Robbins, if k is large and $N(X)$ denotes the number of observations which are equal to X, then the estimator $\hat{\vartheta}_i = (X_i+1)N(X_i+1)/N(X_i)$ is better than $\hat{\vartheta}_i = X_i$. The essence of the paradox is the following: it is possible that observations which have nothing to do with a parameter can influence its good estimations (cf. Paradox II. 2. (*ii*)).

(Ref.: Robbins, H., "An empirical Bayes' approach to statistics", *Proc. 3rd Berkeley Symp. on Math. Statist. and Prob.* I, 157—164, 1956.)

m) A Bayes model paradox

Let the density function $f_p(x)$ of a random variable X be the mixture of two positive density functions $f_0(x)$ and $f_1(x)$:

$$f_p(x) = pf_0(x)+(1-p)f_1(x), \quad \text{where} \quad 0 \leqq p \leqq 1.$$

The value of p is unknown and we hope that we can determine it as precisely as desired if n is large enough, on the basis of the independent observations $X_1, X_2, \ldots, X_n$ (the distributions of X_i's and X are the same). We wish to solve the problem using Bayes' theorem: we choose a number p_0, $0<p_0<1$, and assume that the *a priori* density of X is $p_0 f_0(x)+(1-p_0)f_1(x)$. Then the *a posteriori* density of X (having observed the sample $X_1, X_2, \ldots, X_n$) is: $p_n f_0(x)+(1-p_n)f_1(x)$, where

$$\frac{p_n}{1-p_n} = \frac{p_0}{1-p_0} \prod_{i=1}^{n} \frac{f_0(X_i)}{f_1(X_i)}.$$

The sample elements would really determine the value of p as precisely as we wish if

$$\lim_{n\to\infty} \frac{p_n}{1-p_n} = \frac{p}{1-p} \quad \text{(with probability 1).}$$

This equation, however, does not always hold; if, e.g., the expectation of

$$\log \frac{f_0(X)}{f_1(X)}$$

is 0, then by the *Chung—Fuchs* theorem (cf. III. 7. b.)

$$\lim_{n\to\infty} \sup \frac{p_n}{1-p_n} = \infty \quad \text{and} \quad \lim_{n\to\infty} \inf \frac{p_n}{1-p_n} = 0,$$

therefore

$$\lim \frac{p_n}{1-p_n}$$

does not even exist in this case. The paradox vanishes if instead of $(p_0, 1-p_0)$ we choose an *a priori* distribution which has a positive density function on the whole interval $0<p<1$. This model is more advantageous since it takes the actual f_p into consideration with positive density. (Ref.: Berk, R. H., "Limiting behaviour of the posterior distributions when the model is incorrect", *Annals of Math. Statist.*, **37**, 51—58, (1966).)

n) A paradox of confidence intervals

Let $X_1, X_2, X_3, \ldots$ be normally distributed random variables with a common expectation m and unit variance, and let S_n denote the following sum:

$$S_n = X_1 + X_2 + \ldots + X_n.$$

The probability that for *any fixed n*

$$\frac{S_n - 2\sqrt{n}}{n} < m < \frac{S_n + 2\sqrt{n}}{n}$$

is approximately 95%, whereas the probability that the inequalities hold for *every n* is 0. The latter probability remains 0 even if we substitute an arbitrary large number for 2. (cf. Robbins, H., "Statistical methods related to the law of the iterated logarithm", *Annals of Math. Statist.*, **41**, 1397—1409, (1970).)

o) A paradox of testing independence; is an effective medicine effective?

The three tables below indicate the effect of a certain kind of medicine when it was taken only by men, only by women, and by the two sexes together (combined results). The tables show that the recovery-rates are better after medicinal treatment both among men and women. (The

MEN

	After medical treatment	Without medical treatment
Recovered	700	80
Not recovered	800	130

WOMEN

	After medical treatment	Without medical treatment
Recovered	150	400
Not recovered	70	280

COMBINED

	After medical treatment	Without medical treatment
Recovered	850	480
Not recovered	870	410

Figure 11.

significant difference can be shown statistically by independence tests.) On the other hand, the table of combined results indicates, surprisingly, that the rate of recovery is better among those people who did not take the medicine. So the medicine which proved to be effective both among men and women gave a negative result when a mixed group of men and women were treated with it. Similarly, a newly discovered medicine may be found to be effective in each of ten different hospitals, but the combined list of experiments shows the medicine to be worthless or of negative effect. (Ref.: Pflug, G.: "Paradoxien der Wahrscheinlichkeitsrechnung", in: *Stochastik im Schulunterricht,* Wien, Teubner, 155—163, 1981.)

p) Paradox of computer statistics

The face of statistics has been changed by computers since the 1950's. Without computers scientists were forced to use oversimplified models even if these models were unrealistic. In the last thirty years, however, any statistical decision that a computer can calculate in a relatively short period has become "easy". Thus many "stable" ("robust") and multivariate methods with an enormous quantity of operations entered the practice of everyday statistics. At the same time statistics has become, at least partly, an empirical science: computers can generate millions of data in a few minutes and using them we can "test" most new methods. Many "empirical theorems" were put into practice without firm theoretical basis. On the other hand, the theory of robust statistics (see, e.g., Huber, P. J., *Robust Statistics,* Wiley, New York, 1981) gives the theoretical background for many empirical "dirty tricks" in the practice of statistics. For the controversies and paradoxes of this new period we only refer to the following outstanding papers.

Efron, B., "Bootstrap methods: another look at the jackknife", *Annals of Statist.,* **6**. 1—26, (1979).

Efron, B., "Computers and the theory of statistics: Thinking the unthinkable", *SIAM Review,* 1979 Okt.

Hampel, F. R., "Robust estimation: A condensed partial survey", *Zeitsch, Wahrsch, theorie vrw. Geb.,* **27**, 87—104, (1973),

Miller, R. G., "The jackknife–a review", *Biometrika,* **61**, 1—15, (1974).

Tukey, J. W., "The future of data analysis", *Annals of Math. Statist.,* **33**, 1—67, (1962).

Chapter 3

Paradoxes of random processes

"But next in order I will describe in what ways that assemblage of matter which you see has established earth and sky and the ocean deeps, and the courses of sun and moon. For certainly it was no design of the first-beginnings that led them to place themselves each in its own order with keen intelligence, nor assuredly did they make any bargain what motions each should produce; but because many first-beginnings of things in many ways struck with blows and carried along by their own weight from infinite time unto this present, have been accustomed to move and to meet in all manner of ways, and to try all combinations, whatsoever they could produce by coming together, for this reason it comes to pass, that being spread abroad through a vast time, by attempting every sort of combination and motion, at length those come together which being suddenly brought together often become the beginnings of great things, of sea and sky and the generation of living creatures."

(Lucretius, *De Rerum Natura,* Book V, 416—431, Trans, W.H.D. Rouse)

The first remarkable results in the theory of random processes—or stochastic processes to use a term of Greek origin—arose only in the last century. In the 17th and 18th centuries the chief tendency in investigation was to examine deterministic processes—due especially to the successes of classical mechanics. The "mechanical deterministic" aspect of science, which identified chance with unimportance and aimed to eliminate chance from basic sciences if possible, also evolved at that time. In the second half of the last century, however, the mathematics of random processes gained ground gradually in every fundamental branch of science, also in physics through statistical physics, and played an essential part in 20th century quantum physics. As the profundity of scientific cognition increased, the indispensability of stochastic processes became more and more evident.

1. THE PARADOX OF BRANCHING PROCESSES

a) The history of the paradox

In the first half of the previous century an interesting phenomenon was noticed, namely the gradual extinction of several famous common and aristocratic family names. This problem was studied mathematically by *I. J. Bienaymé* in 1845 and *de Condolle* in 1873. In 1874 *Galton* and *Watson* published a paper of fundamental importance on this subject. The branching chain of family names became the first example of the random branching process. This type of process appeared in chemistry, physics, and in several other areas. E.g., in nuclear physics the process of neutron multiplying or chain reactions can be modelled as branching processes. Neutron generations, however, follow each other much more often than human generations, but in both cases the main question is the same: under what conditions will the process die out (the family name become extinct) or increase to infinity (the bomb blow up). The notion of branching process was coined by *A. N. Kolmogorov* and *N. A. Dmitriev* in 1947.

b) The paradox

Let $p_0, p_1, p_2, \ldots$ denote the probability that an adult man has 0, 1, 2, ... sons. Calculate the probability q that after some generations there remain no male offspring (extinction). Let the generating function of the probability distribution $p_0, p_1, p_2, \ldots$ be defined by

$$g(z) = \sum_{k=0}^{\infty} p_k z^k$$

where $|z| \leqq 1$. Denote the similar generating function in the nth generation by $g_n(z)$. $(g_1(z) = g(z).)$ Then, one can easily see that $g_{n+1}(z) = g(g_n(z))$, i.e., the generating function can be obtained by successive function iterations of $g(x)$. The probability that there remain no male offspring in the nth generation is:

$$q_n = g_n(0).$$

Since q_n is a monotone increasing sequence $\lim q_n = q$ exists and so $q_{n+1} = g(q_n)$ implies

$$q = g(q).$$

Consequently, the probability q can be calculated from this equation. Since $q=1$ is always a root of the equation, Watson supposed (wrongly) that the probability of extinction is always 1 and therefore is unavoidable. Though Watson's result is completely unbelievable, it was not until the 1920s that *R. A. Fisher, J. B. Haldane, J. F. Steffensen* and others showed that the equation has another root, too, which is less than 1 provided that the average number of the sons to be born:

$$m = \sum_k kp_k$$

is greater than 1. In this case the smaller root gives the actual probability of extinction. On the other hand, it is no wonder that in the case when m is less than 1, the probability of extinction is 1. A paradox may only arise in case $m=1$. Supposing that every man has just one son on the average ($m=1$), the probability of extinction is still 1 (except the degenerate case $p_1=1$). Therefore, in spite of the fact that the average number of male offspring remains unchanged over generations (it is always 1) the extinction is unavoidable (more precisely its probability is 1), though one can show that the expected time passing till the extinction is infinite.

c) The explanation of the paradox

The equations

$$q = \lim_{n \to \infty} q_n = 1 \quad \text{and} \quad m = 1$$

do not contradict each other. The first equation means that the probability of a male baby is nearly 0 in the nth generation, but if there are some, then their number may be large, so the average can easily be 1.

d) Remarks

The Galton—Watson model is generally used in the special form when $p_k=ab^{k-1}$ $(k=1, 2, ...)$ and $p_0=1-p_1-p_2-...$ where a and b are positive numbers and a is less than $1-b$. In this case $g_n(z)$ is a simple quotient of linear functions. In 1931 *A. J. Lotka* calculated the above values concerning the USA. He obtained $a=0.2126, b=0.5893$ and $p_0=$ $=0.4825$, so the probability of the extinction of a male line was $q=$ $=0.819$. Nice old family names are gradually becoming extinct and being replaced by more common dull ones like Smith etc. Even the use of combination of two or three names is not always enough to avoid finding identical names, even in one office. The following genetical type naming would be challenging, fair and symmetrical between the two sexes. Each child would inherit two family names, one from the mother and one from the father. Since both parents would also have two family names, they could select their less common (or more attractive) names for the child. Besides these two family names they would, of course, have the first name (or names) as well. Due to this method, our world of names would become more colourful and characteristic.

e) References

Bienaymé, I. J., "De la loi de multiplication et la durée des familles", *Soc. Philomath, Paris,* 37—39, (1845).

Harris, T. E., *The Theory of Branching Processes,* Springer, Berlin—Göttingen—Heildelberg, 1963.

Jagers, P., *Branching Processes with Biological Applications,* London, Wiley, 1975.

Lotka, A. J., "The extinction of families I—II", *J. Wash. Acad. Sci.,* **21**, 377—380 and 453—459, (1931).

Schrödinger, E., "Probability problems in nuclear chemistry", *Proc. Roy. Irish Acad.,* **51**, (1945).

Watson, H. W., Galton, F., "On the probability of the extinction of families", *J. Anthropol.* Inst. Great Britain and Ireland, **4**, 138—144, (1874).

2. MARKOV CHAINS AND A PHYSICAL PARADOX

a) *The history of the paradox*

The concept of Markov chains is due to *A. A. Markov,* a Russian mathematician, whose first paper on this topic was published in The Notes of the Imperial Academy of Sciences of St. Petersburg in 1907. He applied this new concept to the study of the statistical behaviour of the letters in Onegin, the famous poem by Pushkin. The notion "Markov chain" is the most important mathematical notion that originated (at least partly) from linguistics. A sequence (chain) of discrete-valued random variables $X_1, X_2, \ldots, X_t, \ldots$ is called a Markov chain (by definition), if for any initial time t the future (after-t) "behaviour" of the sequence depends on the past (before-t) "behaviour" only through the value X_t, i.e.,

$$P(X_{t+1} = i_{t+1} | X_t = i_t, X_{t-1} = i_{t-1}, \ldots) =$$

$$= P(X_{t+1} = i_{t+1} | X_t = i_t)$$

holds for every possible values $i_{t+1}, i_t, \ldots$ of the random variables, that is, for every possible state. This type of sequence occurs in many fields, e.g., in classical physics, where the future development of a system is completely determined by its present state (e.g., by the instantaneous velocity and position), independently of the way in which the present state has developed. If $\{X_t\}$ is a Markov chain and the conditional probabilities $P(X_{t+1}=i_{t+1}|X_t=i_t)$, the *transition probabilities,* are independent of t, then the Markov chain is called homogeneous. The transition probabilities of homogeneous Markov chains can be arranged in a matrix $A=(p_{ij})$, where

$$p_{ij} = P(X_{t+1} = j | X_t = i).$$

The nth power of this transition matrix is $A^n=(p_{ij}^{(n)})$, where $p_{ij}^{(n)}= =P(X_{t+n}=j|X_t=i)$. This relation allows us to utilize matrix-theory in the theory of Markov chains. Nowadays Markov chains (and their generalization for continuous time parameter and continuous phase space, the Markov processes) are much more important for natural and technical sciences than for linguistics, where they were originally applied.

The problem of reversibility-irreversibility is an interesting paradox of classical mechanics and thermodynamics and Markov chains are efficient means of studying it. The essence of the problem is that the laws of classical mechanics are reversible, so they cannot explain why a cube of sugar dissolves in coffee and why we have never observed the reverse process. The second law of thermodynamics, however, (which was first formulated by *L. S. Carnot*) expresses the irreversibility of our world. (The first law of thermodynamics expresses the principle of conservation of energy.) Forty years later *R. Clausius* introduced the mathematical form of entropy, which is fundamental in the theory of irreversible processes. (According to Clausius, [Memoir read at the Philos. Soc. Zürich, April 24. Pogg. Ann. 125:353, 1865] the word "entropy" comes from the Greek *τϱoπη*, meaning "a turning", or "a turning point". Clausius also states that he added the "en" only to make the word sound like "energy", though the word *εντϱoπη* itself has a meaning, namely "to turn one's head aside".) By means of entropy the second law of thermodynamics can be formulated as follows: in the case of an isolated system the entropy can never decrease, usually it tends to increase. This law was aimed to verify by *L. Boltzmann* using the kinematics of atoms and molecules. (At that time Boltzmann's idea was not natural at all since many physicists doubted even the existence of atoms, e.g., *M. Faraday, E. Mach* or *W. F. Ostwald,* who was the founder of energetics.) Boltzmann was strongly influenced by Maxwell's work on the dynamic theory of gases. In the 1870s Boltzmann found the connection between entropy and thermodynamical probability (cf. Remark (*i*)). He showed that irreversibility does not contradict Newton's reversible mechanics: applying the latter to a large number of particles it necessarily leads to irreversibility, since systems consisting of millions of molecules tend toward a state of greater thermodynamical probability. This is the "final reason" for disintegration, amortization, aging (and moral or historical decay as some say).

In 1907 *P.* and *T. Ehrenfest* created a model which elucidates the paradox of reversibility-irreversibility by the help of Markov chains.

b) The paradox

Suppose we have a system of N molecules; each of them can be in one of two possible energy levels (states). If a molecule is in the first state, in one step it will get into the other state with probability p (and stays in the first one with probability $1-p$); if it is in the second state, (in one step) it moves to the first one with probability q (and remains in the second state with probability $1-q$). As each molecule can "choose" from two possible levels, the system of N molecules can be in 2^N different states. If we consider the molecules indistinguishable, the system can be only in $N+1$ different states: the state of the system is determined by the number of molecules on the first level. Let X_t denote the (random) number of molecules being in the first state at time t. Then $X_1, X_2, X_3, \ldots$ is obviously a Markov chain, which describes the development of the system. How can this model reconcile the reversibility of classical mechanics (symmetry in time) and the irreversibility of thermodynamics (asymmetry in time)?

c) The explanation of the paradox

It can be shown that if $|1-p-q|<1$, then the limit of the generating function of the distribution $P(X_t=j, X_{t+s}=k)$ is

$$\lim_{t\to\infty} E(z^{X_t} w^{X_{t+s}}) =$$

$$= \left[\frac{(p+qz)(p+qw)+pq(1-p-q)^s(1-z)(1-w)}{(p+q)^2}\right]^N,$$

and this function is symmetric in z and w, therefore in equilibrium:

$$P(X_t = j, X_{t+s} = k) = P(X_t = k, X_{t+s} = j).$$

This equation expresses the symmetry between the past and future of the process (reversibility), whereas the following relation expresses irreversibility:

$$P(X_{t+s} = k | X_t = j) \neq P(X_{t+s} = j | X_t = k).$$

If, for example, $p=q=1/2$, then

$$\lim_{t\to\infty} P(X_t = j) = \binom{N}{j} 2^{-N},$$

so the probability that the Markov chain approaches $N/2$ is greater than the probability that it moves away from $N/2$.

d) Remarks

(*i*) Let $f(v, t)$ denote the distribution of the random velocities of gas molecules at time t (for simplicity we assume that the distribution is independent of the position of the gas molecules). Boltzmann formulated his theorem in the following way in 1872: the derivative of the function

$$H(t) = \int f(v, t) \log_a f(v, t)\, dv, \quad a > 1,$$

cannot be positive, that is, H cannot increase as t increases. ($-H$ corresponds to thermodynamical entropy, which, accordingly, may not decrease —it usually increases.) In 1876. *J. Loschmidt,* an Austrian physicist, raised the question of reversibility-irreversibility in the following form: the laws of classical physics are invariant under the transformation $t\to -t$ (they contain second derivates with respect to t), whereas the transformation $t\to -t$ turns Boltzmann's theory to the opposite: $H(-t)$ can never decrease. Through the analysis of this paradox it turned out that for the proof of Boltzmann's theorem the perfect homogeneity of molecular collisions has to be assumed, which is an exaggerated idealization. Boltzmann's theorem is valid only statistically: the probability that $H(t)$ increases as time passes is very small.

Another paradox followed from the theorem of *H. Poincaré.* He showed that considering a closed and finite gas system, the phase point, which describes the state of this system (and moves on an equipotential surface of the multidimensional Euclidean space) returns to an arbitrary small neighbourhood of its initial position within finite time. But this— as *E. Zermelo* showed in 1896—contradicts Boltzmann's theorem: if a process is irreversible (its entropy increases), its phase point cannot be recurrent. The statistical formulation of Boltzmann's theorem, however, solves this problem, too: a sequence of events with very small probability may lead to the return of the phase point, but, according to Boltzmann, it takes $10^{10^{19}}$ years, so this event is practically unobservable, whereas irreversibility can easily be observed.

The Loschmidt—Zermelo paradoxes show that probability theory is a crucially important part of the foundation of molecular physics. Hungarian physicists also gained distinction in this foundation. For example, in 1926 *Leo Szilárd* buried the Maxwell demon which gleamed with the possibility of a "perpetual motion" machine. (Maxwell stated that if the increase of entropy is only statistical, a demon who can track the motion of every molecule, could make a perpetual motion machine. But, according to Szilárd, such a "well-informed" demon requires great entropy, therefore a perpetual motion machine directed by the Maxwell-demon cannot be realized.)

(*ii*) The statistical analysis of the text of Onegin was not an isolated research at all. At the end of the last century it became fashionable to examine the frequency distribution of words in different texts (to help language teaching and shorthand writing). The first frequency dictionary was published in 1898 by *F. W. Kaedig* (Häufigkeitswörterbuch der Deutschen Sprache), and it was based on a text consisting of 11 million words. The application of mathematical statistics in linguistics, however, has become a separate science owing especially to the American scientist *G. K. Zipf* (1902—1950). His book "Human behavior and the Principle of Least Effort" expounds a very complex idea.

e) References

Boltzmann, L., *Lectures on Gas Theory*, Univ. California Press, Berkeley, 1964.

Chung, K. L., *Markov Chains with Stationary Transition Probabilities*, Springer, Berlin—Göttingen—Heidelberg, 1960.

Dynkin, E. B., *Markov Processes*, Springer, New York, 1965.

Ehrenfest, P., Ehrenfest, T., "Über zwei bekannte Einwände gegen das Boltzmannsche H-Theorem", *Physik. Zeit.*, **8**, 311—314, (1907).

Feller, W., *An Introduction to Probability Theory and Its Applications*, New York, John Wiley, 1969.

Kempermann, J. H. B., *The Passage Problem for a Stationary Markov Chain*, Univ. Chicago Press, 1961.

Markov, A. A., "Extension of the limit theorems of probability theory to a sum of variables connected in a chain", *The Notes of the Imperial Academy of Sciences of St. Petersburg,* VIII. Ser. Phys.-Mat. Collage, XXII, Dec., **5**, (1907).
Takács, L., "On an urn problem of Paul and Tatiana Ehrenfest", *Math. Proc. Camb. Phil. Soc.,* **86**, 127—130, (1979).
Truesdell, C., *The Tragicomic History of Thermodynamics 1822—1854,* Springer, New York, 1980.

3. THE PARADOX OF BROWNIAN MOTION

a) The history of the paradox

While performing microscopic experiments *Robert Brown,* the English botanist (1773—1858), discovered not only the nucleus of the cell but also another interesting though at that time unexplainable phenomenon: the random motion of colloid-size particles, known today as Brownian motion. Since he made his first experiments (June—August 1827) with pollen, it was supposed that the motion was biological. Brown's great merit was the experimental proof of the sole physical nature of the phenomenon. At that time microphysics was not developed enough to be able to explain the phenomenon. No wonder that even in 1879 *C. W. Nägeli,* a Swiss—German biologist, refused to believe that the Brownian motion was due to the thermal diffusion of molecules. On the other hand, *J. H. Poincaré* claimed in a lecture (Paris, 1904) that when big particles about the size of 0.1 mm are hit very many times from all directions by moving atoms, they do not move because the random collisions neutralize each other, according to the laws of large numbers; smaller particles, however, do not get pushed enough to neutralize each other so the particles move in a zigzag path. The quantitative explanation was given independently by *Einstein* and the Polish *Smoluchowski* in 1905. According to Einstein's theorem, the mean path of the particles is proportional to the square root of the time t. Consequently, their mean speed is proportional to $1/\sqrt{t}$. From this it follows that the instantaneous speed of the particles would be infinite in any moment showing that there are problems in defining the instantaneous speed for the Brownian motion. A deeper mathematical analysis was required to solve this problem. It was not performed until more than a decade later by *N. Wiener*. To acknowledge his merits in this field, the mathematical model of Brownian motion is called after him: the Wiener process. The Wiener process is a motion with continuous path (its realizations are continuous) which is nowhere differentiable with probability 1. This means that the instantaneous speed cannot be defined anywhere.

Functions which are everywhere continuous but nowhere differentiable were already known by mathematicians long before Wiener. Such pathological functions where, however, considered only curiosities. In 1806 *A. M. Ampère,* the famous physicist, even wanted to show that, apart from some isolated points, every continuous function is differentiable. Due to researches mainly on Fourier series, the notion of function was made much more general by *B. Bolzano* (1834), *G. F. B. Riemann* (1854) and *K. Weierstrass* (1872). Weierstrass' continuous but nowhere differentiable function was firstly published by *P. Du. Bois-Reymond* in 1875. Most outstanding mathematicians were not too enthusiastic about this invention. According to Poincaré (Science et Méthode, 1909), "In the old days when people invented a new function they had some useful purpose in mind: now they invent them deliberately just to invalidate our ancestors' reasoning, and that is all they are ever going to get out of them." *Ch. Hermite* wrote to *I. J. Stieltjes* in a similar way. "With horror and dread do I turn away from this miserable plague: functions that have no derivatives." The Wiener process was an obvious refutation of the above accusations, because nobody could say that the Brownian motion was invented only to create a pathological counterexample. 20th century researches also made it clear that among continuous functions just the undifferentiable ones are typical, in a sense, they are in overwhelming majority. (Oxtoby, L. J. C., *Measure and Category*, Springer, New York, 1971). In practice, however, most continuous functions are differentiable. It is just like the case of irrational numbers. In spite of their majority among real numbers (a random number is irrational with probability 1), in practice we generally use rational numbers.

b) The paradox

The trajectories (realizations) of Brownian motion are rather irregular (i.e., they are nowhere differentiable). In the usual sense we consider any irregular curve, such as the trajectory of planar Brownian motion, one dimensional. At the same time it can be shown that the trajectory of a planar Brownian motion actually fills the whole plane (each point of the plane is approached with any given accuracy with probability 1). Therefore the trajectories can also be considered as two-dimensional curves. Which conception is preferable?

c) The explanation of the paradox

The notion of dimension was used in the common, every day sense even at the beginning of this century. Curves, surfaces, and bodies were considered one, two, and three-dimensional, respectively. Generally, a figure is said to be k-dimensional if k parameters (coordinates) are required to "characterize" the points of it. Using Poincaré's intuitive ideas *L. E. J. Brouwer* defined the topological dimension in 1913. Later, in 1922, *K. Menger* and *P. S. Uryson*, working independently, also succeeded in defining it. (For more details see the book by Hurewitz and Wallman.) By the definition of topological dimension, the Brownian motion is one-dimensional. On the other hand, in 1919 *F. Hausdorff* introduced the following notion of dimension, according to which the Brownian motion is two-dimensional. In the d-dimensional Euclidean space the volume of the unit sphere is $v(d)=\Gamma(1/2)^d/\Gamma(1+d/2)$, where Γ denotes the usual gamma function (see the Notations). This expression has sense even if $d \geqq 0$ is not integer. Let a set E be given in the n-dimensional Euclidean space, which is covered by a finite number of n-dimensional spheres with radii $r_1, r_2, \ldots$. The Hausdorff d-measure of the set E is then

$$\lim_{r\to 0} \inf_{r_i<r} \sum_i v(d) r_i^d.$$

A. S. Besicovitch has proved that there always exists a (real) number D that in case $d<D$ the d-measure of the set E is infinite but if $d>D$ it is 0. This number (D) is called the Hausdorff or Hausdorff—Besicovitch dimension of the set E. In this sense the value of the dimension need not be an integer. E.g., both coordinates of the planar Brownian motion as functions of time (i.e., the curves of the "one-dimensional Brownian motion") have Hausdorff-dimension of 3/2. These curves are therefore somewhere between being a "real" curve and a "real" surface. The dimension of the curve of the planar Brownian motion is 2 just like that of the "real" surfaces.

d) Remarks

(*i*) In the last few years many papers have been published on figures whose topological and Hausdorff-dimensions are different. *B. Mandelbroit* called them *fractals*. Fractals, e.g., Wiener processes, play a fundamental role in describing irregular figures of nature. While the Euclidean line is the most frequent "letter" describing regular forms of nature, for irregular forms (clouds, seashores) it is the Wiener process. In fact neither "real" lines (having extension in length only) nor "real" Wiener processes (nowhere differentiable) exist in nature but with their help a fairly good picture of "real" forms can be obtained. Fractals have also put the famous *Olbers'* paradox of astronomy in a new light. According to the paradox, it is inconceivable that the sky does not shine uniformly at night if the stars are uniformly distributed in space. (See Mandelbroit's book.)

(*ii*) In his book Mandelbroit also mentions other notions of dimension, such as the *Fourier* dimension. As for algebraic dimension see *Székely*'s article.

(*iii*) The irregularity of the Wiener process led to the development of a new frontier of probability theory and analysis, namely the theory of stochastic differential equations. This theory produces great deviations from the usual differential and integral calculus. E.g., if $f(t)$ is a differentiable function, then

$$\int_0^1 f(t)\, df(t) = \int_0^1 f(t) f'(t)\, dt = \frac{f(1)^2 - f(0)^2}{2}.$$

In the theory of stochastic integrals the above expression makes sense even if $f(t)$ is the (nowhere differentiable) Wiener process. In this case the result of the integral is less than in the above (differentiable) case. The difference is exactly 1/2.

e) References

Bachelier, L., *Théorie de la spéculation*, Thesis, 1900.

Brown, R., "A brief account of microscopical observations made in the months of June, July and August 1827, on the particles contained in the pollen of plants; and on the general existence of active molecules in organic and inorganic bodies", *Edinburgh New Phil. Journal*, **5**, 358—371, (1828).

Gihman, I. I., Skorohod, A. V., *Introduction to the Theory of Random Processes*, Moscow, Nauka, 1965 (in Russian).

Hausdorff, F., "Dimension und äusseres Mass", *Math. Annalen*, **79**, 157—179, (1919).

Hurewitz, W., Wallmann, H., *Dimension Theory*, Princeton Univ. Press, 1941.

McKean, H. P., Jr., *Stochastic Integrals*, Academic Press, New York—London, 1969.

Mandelbroit, B. B., *Fractals, From, Chance and Dimension*, W. H. Freeman and Co., San Francisco, 1977.

Székely, G. J., "Algebraic dimension of semigroups with application to invariant measures", *Semigroup Forum*, **17**, 185—187, (1979).

Wiener, N., *Collected Works*, Cambridge, Mass. M. I. T. Press, 1976.

4. THE PARADOX OF WAITING TIMES (DO BUSES RUN MORE FREQUENTLY IN THE OPPOSITE DIRECTION?)

a) The history of the paradox

Though modern technology continually shortens wasted waiting times, our everyday nervousness is in great part due to useless waits. So the efforts of mathematicians and engineers to reduce waiting times is followed with great interest. It was *A. K. Erlang* who examined waiting time problems for telephone exchanges (cf. I/6 Remark (*iii*)). In the 1930s *W. Feller* introduced the notion of birth and death processes, which gave an impulse to the mathematical analysis of waiting and greatly contributed to the emerging theory of operations research. The study of queuing systems has become an independent branch of science on the borderland between probability theory and operations research.

b) The paradox

The "frequency of bus runs", i.e., average time passing between the arrival of two consecutive buses is usually indicated at the bus stops. Suppose the frequency of runs at a certain bus stop is 10 minutes. Then we expect that people have to wait 5 minutes on the average for a bus. It was found, however, that the average waiting time may not only exceed 5 minutes, it may even be infinite! (Experience shows, however, the situation in everyday life is not so bad.)

c) The explanation of the paradox

If buses arrive at the bus stop not only on an average but exactly every 10 minutes, the average waiting time would really be 5 minutes. But buses actually run in "packs" (except if they are near the terminal where they started from). Therefore waiting times show a large dispersion about the average value. Let us suppose that the time intervals between consecutive arrivals of buses at a bus stop are independent identically distributed random variables with expectation m and standard deviation s. Then it can be shown that the average waiting time is $T=(m^2+s^2)/2m$. Let $F(t)$ be the distribution function and $f(t)$ be the density function of the intervals between consecutive arrivals of buses. (We assume now that the density function exists, though this condition may be omitted at the cost of some modifications.) Suppose that time t is measured from the arrival of the last bus before our arrival. Then the density function of the random time interval till the arrival of the next bus is not $f(t)$ but another function which is proportional to $t \cdot f(t)$ (i.e., $t \cdot f(t)/m$), since the probability that we arrive during a certain time interval is proportional to its length t. Thus the average waiting time is.

$$T = \frac{1}{2m} \int_0^\infty t^2 f(t)\, dt = \frac{m^2+s^2}{2m}.$$

(The density function of our waiting time is $(1-F(t))/m$.) Accordingly, $T=m/2$ only if $s=0$, but if $s=\infty$ then $T=\infty$, too. These extremes are naturally far from reality. Buses actually arrive at intervals which have nearly exponential ("ageless") distribution with some parameter λ.

Then $m=s=1/\lambda$, that is, $T=m$, meaning that if the frequency of runs is 10 minutes, the average waiting time is also 10 minutes and not 5.

The heuristic explanation of this paradox is very simple. If somebody arrives at the bus stop at random, he has a much greater chance of waiting for a long time, since his waiting time would only be short if he caught one of the buses in a "pack"; but the buses of a "pack" arrive at very short intervals so one has not much chance of catching any of them. Consequently, if the time intervals between consecutive buses show great dispersion, there are only a few people who have a short wait and there are many people who have a long wait, meaning that the average waiting time T is large.

d) Remarks

(*i*) We often have the illusion that no matter which way we would like to go, buses and trams run more frequently in the opposite direction. Naturally this is impossible in reality. The explanation is very simple. We see only one bus (the one we take) which runs in the same direction as we want to go, whereas the probability of two or three buses passing in the opposite direction while we are waiting is positive. Their expected number is

$$\frac{m^2+s^2}{2m} : m = \frac{1}{2} + \frac{s^2}{2m^2},$$

which is really greater than 1/2 if s is positive. This shows an asymmetry between the two directions. But in fact this is not the case. The symmetry between the two directions can be expressed by the fact that the probability that no bus will go in the opposite direction while we are waiting for our bus is just 1/2 (but if a bus goes in the opposite direction then more than one may also go, so the expectation may be arbitrarily large). Let p_k denote the probability that exactly k buses pass in the opposite direction while we are waiting. If the intervals between the arrivals of consecutive buses are exponentially distributed, then

$$p_k = \left(\frac{1}{2}\right)^{k+1};$$

if they are uniformly distributed on the interval (0, 1), then

$$p_k = 4\left(\frac{1}{(k+2)!} - \frac{2}{(k+3)!} + \frac{1}{(k+4)!}\right),$$

where $k=1, 2, \ldots$ (p_0 is always equal to 1/2 as we have mentioned).

(*ii*) Where there is a congestion of bus traffic, the bus service would become steadier if the congested buses waited for longer time at a bus stop; thus the average waiting time would also shorten. (Actually I have never seen buses waiting at a bus stop, just to make the traffic steadier, though lifts are sometimes held back to wait for people who are likely to arrive soon. So lift traffic is slowed down to shorten the average waiting times!) Let $t_1, t_2, t_3, \ldots$ denote the times when buses arrive to a certain bus stop, and let $X_1 = t_2, X_i = t_i - t_{i-1}$, $(i=2, 3, \ldots)$. If the distribution function of $X_1, X_2, \ldots$ is F, then—as we have already mentioned—the density function of X_1 is $[1-F(t)]/m$, whose expectation is $T=(m^2+s^2)/2m$. Slowing down the traffic means that we increase the X_i's to $X_i + g(X_i)$, $(i=2, 3, \ldots)$ by a non-negative function g. It can be shown that among integrable functions the function $g(x) = \max(0, (c-x))$ most shortens waiting times, where c is the unique solution of the following equation:

$$cE(X) + \int_0^c (c-x)F(x)\,dx = E(X^2)/2,$$

(where X is a random variable having the same distribution as $X_2, X_3, \ldots$. If for example X is exponentially distributed, more precisely if $F(x) = 1-e^{-x}$ $(x>0)$, then both the expectation and the variance of the waiting time equals 1. If we choose the optimal $g(x) = \max(0, (0.901-x))$ (accurate to three decimal places) then the expected value of our average waiting time is only 0.901 (and the variance is 0.691).

(*iii*) The following paradox is connected with traffic, too. (*G. Schay* drew my attention to the problem itself, after my talk in MIT in 1983.) The paradox states that it is not true that the faster cars go the more of them can get through the green light, as, at higher speeds, cars have to keep greater distances. Let us start from the following model to calculate the optimal speed. Suppose cars go at the same speed v; let X_i denote the time (depending on chance), between the ith and the $(i+1)$th car getting into the traffic; X_i's are independent and identically distributed (for simplicity we shall assume that this distribution is exponential with a parameter $\lambda > 0$). Cars arrive at the traffic lights at intervals $Y_1, Y_2, \ldots$ which are not simply equal to $X_1, X_2, \ldots$, since cars have to follow each other at a certain distance. Let l_i denote the length of the ith car and a_i its braking deceleration; l_i and a_i are usually not independent, but we assume that the vectors (l_i, a_i) $i=1, 2, \ldots$ are independent and identically distributed. The braking distance is $v^2/2a_i$, thus cars have to follow each other at the distance $l_i + v^2/2a_i$. The time between the arrival of the first and $(n+1)$th car is

$$\sum_{i=1}^{n} Y_i = \max\left\{\sum_{i=1}^{n} X_i,\ \sum_{i=1}^{n-1} Y_i + Z_{n-1}\right\},$$

where

$$Z_i = \left(l_i + \frac{v^2}{2a_i}\right)\Big/ v.$$

If

$$M(t) = \max\left\{n:\ \sum_{i=1}^{n} Y_i \leqq t\right\},$$

then the number of cars which can get through the green light from time t to $t+h$ is $M(t+h) - M(t)$. It is known (from the theory of queues) that

$$\lim_{t\to\infty} \frac{M(t)}{t} = \begin{cases} E(Z_1)^{-1} & \text{if}\quad E(X_1) \leqq E(Z_1) \\ \text{(this corresponds to the traffic in rush hours)} & \\ E(X_1)^{-1} & \text{if}\quad E(X_1) > E(Z_1). \end{cases}$$

Let t be a random time within the interval $[0, T]$. Then the average number of cars which go through the green light is

$$\frac{1}{T}\int_0^T [M(t+h) - M(t)]\, dt = \frac{1}{T}\left[\int_T^{T+h} M(t)\, dt - \int_0^h M(t)\, dt\right] =$$

$$= h \min\{E(Z_1)^{-1}, E(X_1)^{-1}\} + o(1), \quad \text{if}\quad T \to \infty;$$

($o(1)$ denotes a quantity which converges to 0, as $T \to \infty$.) Therefore we seek the maximum of

$$\min\{\lambda, [E(l_1)/v + E(a_1^{-1})v/2]^{-1}\}$$

for v. The second term is maximal if

$$v = \sqrt{2E(l_1)/E(a_1^{-1})}.$$

e) References

Gadžiev, A. G., "Minimization of the mean waiting time in the system with recurrent service", *Vestnik MGU*, Ser. 1, **3**, 19—24 (in Russian), (1980).
Kleinrock, L., *Queuing Systems*, Wiley, New York, 1975.
Takács, L., *Introduction to the Theory of Queues*, Oxford Univ. Press, New York, 1962.

5. THE PARADOX OF RANDOM WALKS

a) The history of the paradox

About 60 years ago *George Polya*, the American mathematician of Hungarian origin, used to walk in a park where he kept meeting the same couple. At that time he did not realize how accidental these random meetings were, i.e., how small the probability was. Shortly afterwards he calculated the probability of meetings in a model where 2 persons are walking randomly on a squared network independently of each other (at each crossing the probability of choosing any of the four possible directions is the same). Polya found that the probability of meeting was 1. (Consequently if their time were unlimited they could also meet infinitely many times with probability 1.) In the case of a cubic network, however, the probability of a meeting is strictly less than 1 (so the probability of infinitely many meetings is now 0). From this interesting discovery a brand new branch of probability theory has developed during the last 60 years. In 1964 a nice monograph was written on this theme by *F. Spitzer*.

b) The paradox

From Polya's theorem it follows that considering a random walk on the integer points of the real line starting from the origin and moving at every step by 1 either to the left or to the right with the same probability 1/2 (independently of the previous steps), we shall get back to 0 with probability 1. Now the question arises that before returning to 0 (for the first time) how many times has the walk reached a fixed integer k? It is natural to suppose that the greater $|k|$ is, i.e., the farther the random walk goes from the origin the fewer times it will happen, on average. Surprisingly, the random walk will always reach k before the first return just as many times on average, namely once, however great $|k|$ is.

c) The explanation of the paradox

The paradox can be explained very simply. The average number of steps necessary to return to the origin (i.e., the expected value of the recurrence time) is infinite, consequently there is enough time to reach any point on the line once, on average. A related paradox is the following. The starting point of the random walk is only finitely many times is its most visited site (with probability one).

d) Remarks

(*i*) Under the above conditions suppose that we always take 2 steps to the right but only 1 to the left. In this case the random walk is not symmetric and one can easily see that starting from 0 the probability of reaching -1 is less than 1. This probability is, suprisingly, just $(\sqrt{5}-1)/2$, i.e., the ratio of golden section.

(*ii*) Diffusion type random walks where the probability of moving to the left or to the right depends on actual location (k) are very important in practice. Let p_k denote the probability of moving to the right and $1-p_k$ the probability of moving to the left. Suppose furthermore that

$$p_k = \frac{1}{2}\left(1+\frac{c}{k}\right)$$

(at least for great values of $|k|$) where c is an arbitrary constant. This kind of random walk returns to 0 with probability 1 (i.e., it is recurrent) if $c \leqq 1/2$. In the case $c < -1/2$, the expected value of the recurrence time is finite, therefore the paradoxical situation of the random walk (corresponding to $c=0$) cannot appear.

(*iii*) Researches on random walks can be extended from squared networks to more general ones called graphs. These generalizations have interesting applications in the theory of electronic networks. See the fundamental paper by *C. Nash-Williams* written in 1959. Many other applications (in physics, chemistry and biology) are discussed in Weiss' paper.

e) References

Bass, R. F., Griffin, P. S., "The most visited site of Brownian motion and simple random walk", *Zeitsch 'Wahrsch' theorie verw. Geb.* **70**, 417—436, (1985).

Nash-Williams, C. St. J. A., "Random walk and electronic currents in networks", *Proc. Camb. Phil. Soc.*, **55**, 181—194, (1959).

Polya, G., "Über eine Aufgabe der Wahrscheinlichkeitsrechnung und das Momentproblem", *Math. Annalen*, **84**, 149—160, (1921).

Spitzer, F., *Principles of Random Walk*, Van Nostrand, New York—Toronto—London, 1964.

Weiss, G. H., "Random walks and their applications", *American Scientist*, **71**, 65—71, (1983).

6. THE PARADOX OF STOCK EXCHANGE; MARTINGALES

a) The history of the paradox

The mathematical study of the Stock Exchange is of almost the same age as the Stock Exchange itself. Presumably not even *Gresham's* Exchange in the 16th century was free from mathematical speculation, but the basic methods of probability theory were not applied in this field for quite a long time. It is typical that even in 1900, when *Louis Bachelier* defended his doctoral thesis in Paris on the connection between price fluctuations in the Stock Exchange and Brownian motion (preceding the physicists' investigations concerning Brownian motion), the commettee hardly appreciated his essentially new ideas. Bachelier created the general mathematical model of a fair game, the so-called *martingale,* which later became one of the most important stochastic processes after the researches of *J. Ville, P. Lévy, D. L. Doob* and others. A sequence of random variables $X_1, X_2, X_3, \dots$ is called a martingale if the conditional expectation of the difference $X_{n+1}-X_n$ ("the profit gained at time n"), given the total capitals $X_n, X_{n-1}, \ldots$, is zero with probability one, for every n, that is,

$$E(X_{n+1}-X_n|X_n, X_{n-1} \ldots) = 0$$

with probability one. The sequence $X_1, X_2, X_3, \ldots$ is a *supermartingale* (or submartingale) if the above mentioned expectation is not positive (or negative) with probability one. The martingale is a general model of fair game, of "quantitative justice", which can be applied in many fields, for instance, in the study of paradoxes in the Stock Exchange.

b) The paradox

If a share is expected to be profitable, it seems natural that the share is worth buying, and if it is not profitable, it is worth selling. It also seems natural to spend all one's money on shares which are expected to be the most profitable ones. Though this is true, in practice other strategies are followed, because while the expected value of our money may increase (our expected total capital tends to infinity), our fortune itself tends to zero with probability one. So in Stock Exchange business we have to be careful: shares which are expected to be profitable are sometimes worth selling.

c) The explanation of the paradox

Let us suppose that we would like to buy shares, and we can choose from k different ones; in a one year period the ith share ($i=1, 2, \ldots, k$) yields $X^{(i)}$ times as much profit as our initial capital was at the beginning of the year. (Obviously $X^{(i)} \geqq -1$.) Suppose, for simplicity, $X^{(i)}$ is bounded, though this condition can be omitted after some modifications of the reasoning below. The random vector $X=(X^{(1)}, \ldots, X^{(k)})$ describes the quotations. We assume that the vectors X_j ($j=1, 2, \ldots$), which describe the quotations in the jth year are of the same distribution as X and are independent. Let T_0 be our initial capital and let $a_j^{(i)}$ denote the proportion of our total capital that we spent on buying shares of type i in the jth year. The quantity $a_j^{(i)} \geqq 0$ may depend on the random vectors $X_1, X_2, \ldots, X_{j-1}$. The vector $a_j=(a_j^{(1)}, a_j^{(2)}, \ldots, a_j^{(k)})$ describes our buying strategy in the jth year. Evidently

$$\sum_{i=1}^{k} a_j^{(i)} \leqq 1.$$

Let $a_j X_j$ denote the following sum:

$$\sum_{i=1}^{k} a_j^{(i)} X_j^{(i)}.$$

Using this notation, our total capital at the end of the nth year is

$$T_n = T_0 \prod_{j=1}^{n} (1+a_j X_j).$$

Obviously, the expectation of T_n is the greatest if we spend all our money on the most profitable shares every year. (We suppose that at least one of the shares is profitable.) In this case the expected value of T_n tends to infinity (so we are likely to grow rich) and still our total capital T_n may tend to zero with probability one! Let us examine this paradoxical situation in detail. Clearly,

$$\log T_n - \log T_0 = \sum_{j=1}^{n} \log (1+a_j X_j).$$

Assuming that $a_j = a$ is a constant vector (independent of j; this assumption is quite natural since in our case the quotation distributions do not change), the right side of the equation is the greatest (with probability one for large n, according to the law of large numbers) if

$$E\big(\log (1+a_j X_j)\big)$$

is maximal (under the conditions $a_j^{(i)} \geqq 0$ and $\sum_{i=1}^{k} a_j^{(i)} \leqq 1$). Let a^* denote the strategy which maximizes the above quantity. Further let T_n^* and T_n denote, respectively, our total capital if we follow the strategy a^* or an arbitrary strategy a. Then it can be shown that the sequence T_n/T_n^* $(n=1, 2, \ldots)$ is always a non-negative supermartingale (moreover, if every coordinate $a^{*(i)}$ of the vector a^* is positive and

$$\sum_{i=1}^{k} a^{*(i)} < 1,$$

then it is a martingale). Therefore, according to a well-known theorem of martingale theory

$$\lim_{n\to\infty} T_n/T_n^* = T$$

always exists with probability 1 and its expectation is at most 1. Thus a^* is an optimal strategy (in this sense) in the long run. In sum: it is advantageous to maximize the expectation of $\log T_n$ and not that of T_n. The heuristic explanation of this fact is quite simple: T_n increases exponentially for every reasonable strategy, and the rate of this increase can be maximized just by maximizing the expected value of $\log T_n$.

Consider now a simple (but extreme) example. Suppose we can choose from two kinds of shares; $p_{11}=10\%$ is the chance that the values of both shares double, $p_{00}=5\%$ is the probability that both shares lose their value, the probability that the value of the first share doubles and the second deteriorates is $p_{10}=50\%$, and with probability $p_{01}=35\%$ the same happens inversely. Then the first share is profitable (with a probability of 60%) and the second one is losing (it is profitable only with a probability of 45%), but it is still reasonable to buy some of both kinds of shares, more exactly to spend one third of our money on the shares every year in the ratio of 13:4. Generally it is reasonable to spend the

$$\frac{p_{11}-p_{00}}{p_{11}+p_{00}}$$

proportion of our money on the two kinds of shares in the ratio of

$$(p_{11}\,p_{10}-p_{01}\,p_{00}):(p_{11}\,p_{01}-p_{10}\,p_{00})$$

(assuming that the differences are positive). Though the problems which occur in the practice of Stock Exchange business are much more complicated than the preceding example, the paradox in question also appears in these more complex problems.

d) Remarks

(*i*) The martingale as a system of play was well-known long before the appearance of the mathematical theory of martingales. (We shall quote from the paper J. L. Snell; "Gambling, probability and martingales", *The Mathematical Intelligencer,* **4**, 118—124, (1982). "The basic idea of the martingale system is to double when you lose. For example, suppose that we are playing roulette and we bet each time on red. We make an initial bet of $1. If we win we quit; if we lose, we make a bet of $2 next time. If we win, we are $1 ahead and we quit; if we lose, we are down $1 $2 $3, and we bet $4. If we win, we are $1 ahead and quit; if we lose, we bet $8 next time, etc. Under this martingale system, if the wheel ever stops on a red number, we leave the casino $1 richer than when we entered. Since a red is bound to show up eventually, it seems that this is a foolproof system. But suppose that we enter the casino with $100 and we encounter a run of 6 black. Then we have lost $2^6-1=63$ dollars and we cannot make the next required bet of $64.

In his book *Newcomes* (The Newcomes; Memoirs of a Most Respectable Family, Chapter 28, Page 266, London, Bradbury and Evans, 1953), Thackeray remarks "You have not played as yet? Do not do so; above all avoid a martingale if you do." While this is a good advice for the gambler, mathematicians have not heeded it, and many of important results in probability theory have come from ignoring Thackeray's advice.

(*ii*) *Thomas Gresham* (1519—1579) the founder of the London Stock Exchange, must have guessed that mathematics had an important part in the analysis of Stock Exchange and economic life. Gresham's testament included the plan of a college where mathematics was one of the main subjects of economics teaching. *Henry Briggs,* who first published a logarithmic table in 1617, was also a professor at Gresham College, which can be considered, in many respects, the predecessor of the Royal Society.

(*iii*) Martingales have many interesting applications in genetics, potential theory, stochastic integrals, etc. The monographs of *J. Neveu, P. A. Meyer, C. C. Heyde,* and *P. Hall* are outstanding in this field.

e) References

Breiman, L., "Optimal gambling systems for favorable games", *Proc. 4th Berkeley Symp. on Math. Statist. and Prob.*, 65—78, (1961).
Doob, J. L., *Stochastic Processes*, Wiley, New York, 1953.
Móri, T. F., Székely, G. J., "How to win if you can", *Coll. Math. Sci. Bolyai* **36** (ed. P. Révész), 791—806, (1982).
Ville, J., *Étude critique de la notion de collectif*, Gauthier-Villars, Paris, 1939.

7. QUICKIES

a) Jacob and Laban's paradox

According to the biblical story of Jacob and Laban, Jacob got Laban's dappled sheep in return for his services. Though the proportion of dappled sheep to others was very small, Jacob gradually acquired greater wealth than Laban. There are many mystic explanations of this paradox (the Bible itself contains one, and Thomas Mann also dealt with this riddle), but—as Alfred Rényi once pointed out—there is nothing mysterious about this paradox at all; it can be understood by simple mathematical inference based on the fact that Jacob never returned sheep to Laban but Laban always gave Jacob some of his own sheep.

Let us denote the average number of Jacob's and Laban's sheep in the nth year, respectively, by J_n and L_n (in the initial, 0th year $J_0=0$ and L_0 is a positive number). Let us suppose that each sheep has U lambs every year on the average. Let q denote the proportion of Laban's sheep that he gives Jacob ($p=1-q$ proportion remains at Laban). Then $L_{n+1}-L_n=UpL_n$ and $J_{n+1}-J_n=UJ_n+UqL_n$, consequently $L_n=$ $=L_0(1+Up)^n$ and $J_n=L_0(1+U)^n-L_0(1+Up)^n$, therefore

$$\frac{J_n}{L_n}=\left(\frac{1+U}{1+Up}\right)^n-1$$

and this tends to infinity as n increases, so Jacob will really be richer than Laban after a time. For example, for $q=10\%$, $U=2, n=20$, the ratio $\frac{J_n}{L_n}$ is approximately 3.

b) A paradox of processes with independent increments

Processes with independent increments and their discrete versions, the partial sums of independent random variables, are classical areas of probability theory. Let $X_1, X_2, \ldots$ be independent (not identically zero) random variables with expectation zero. Then the sums $S_n = X_1 + X_2 + \ldots$ $\ldots + X_n$, $n = 1, 2, \ldots$ fluctuate about zero, i.e., (according to a theorem of *K. L. Chung* and *W. H. J. Fuchs* proved in 1951) if X_i's have a common distribution, then

$$P(\limsup_{n\to\infty} S_n = +\infty) = P(\liminf_{n\to\infty} S_n = -\infty) = 1.$$

This fluctuating property, however, does not necessarily hold if the X_i's are not identically distributed. Put, e.g., $X_i = Y_i/\sqrt{1-i^{-2}}$, where $P(Y_i = i^{-1}) = 1 - i^{-2}$ and $P(Y_i = -i + i^{-1}) = i^{-2}$. If the Y_i's are independent, then the X_i's are also independent and have expectation zero and variance one. According to the *Borel—Cantelli lemma,* if $A_1, A_2, A_3, \ldots$ are arbitrary events and the sum of their probabilities converges, then, with probability one, only a finite number of events A_k occur. Hence the event $Y_i = -i + i^{-1}$ also occurs only finitely many times (since $\sum_{i=1}^{\infty} i^{-2} < \infty$), so for n sufficiently large, $Y_i = i^{-1}$ with probability one, that is, $X_i = 1/\sqrt{i^2 - 1}$, thus

$$P(\lim_{n\to\infty} S_n = \infty) = 1.$$

c) The paradox of goals

Two teams A and B are playing football against each other. Suppose the teams have equal abilities (i.e., both teams score the next goal with probability 1/2 at any time during the match). If the length of the time interval between two consecutive goals is constant, then it seems natural to think that for 50% of the playing time team A leads and for 50% of the time team B leads. Surprisingly, however, just the contrary is true: it is most improbable that A (or B) will be in the lead for the half of the playing time (if the cumulative scores are equal, the leading team is considered the one which was leading before the last goal). If $n=20$ goals were scored during the match, then the probability that after 10 goals team A and after the other 10 goals team B leads is only 6%; however, the probability that one of the teams will be in the lead throughout the game is approximately 35%. It is also surprising that the probability that one team leads throughout the second half is 50 per cent, no matter how large n is.

The situation changes considerably if the "goal-scoring ability" of the teams depends on the score of the game. Let

$$p_k = \frac{1}{2}\left(1+\frac{c}{k}\right)$$

be the probability that A scores the next goal if team A leads by k goals, and $k \neq 0$; $p_0 = 1/2$. If c is large and k is small, $0 < p_k \leqq 1$ may not hold; then let $p_k = 1/2$. (If $c=0$, then $p_k = 1/2$ for every k and this leads to the simple model we have just examined.) If c is positive, the leading team has more chance of scoring the next goal. If $c > 1/2$, then after a time one of the teams "breaks down", that is, if many goals are kicked, only one team leads (which, depends on chance) nearly during 100% of the playing time. On the other hand, if c is negative, then the losing team scores a goal with greater probability; for $c < -1/2$, the match is very varied and interesting: for half of the play one team leads and for the other half the other team.

It can be shown that for $c=0$, the probability that team A will lead at most for the fraction x $(0 < x < 1)$ of the playing time converges to

$$F(x) = \frac{2}{\pi} \operatorname{arc\,sin} \sqrt{x} \quad \text{as} \quad n \to \infty.$$

The corresponding density function for $0 < x < 1$ is

$$f(x) = \frac{1}{\pi \sqrt{x(1-x)}},$$

which is minimal for $x = 1/2$. Thus the probability density of A leading for exactly 50% of the playing time is really the smallest. This is *Paul Lévy's arc sine law* (1939). (My conjecture is that in the general case the density function is proportional to the $(2c+1)$th power of $f(x)$ if $c < 1/2$.)

Finally one more surprising fact: in the case $c=0$, if the game ends in a tie ($n:n$) and we want to know how long team A was in the lead, and we take the interval between two consecutive goals as the unit of time, then the probability that A was in the lead for a time $2k$ ($k=0, 1, 2, \ldots, n$) is independent of k!

(Ref.: Feller, W., *An Introduction to Probability Theory and its Applications*, John Wiley, New York, 1969.
Lamperti, J., "Criteria for recurrence or transience of stochastic process I," *J. Math. Anal. Appl.*, **1**, 314—330, (1960).)

d) The paradox of expected ruin time

A and B are playing a coin tossing game. If it is heads, then A pays B, if tails, B pays A 1 dollar. A's initial capital is 1 dollar and B's is 999 dollars; they play till one of them is ruined. A has of course more chance of running out of money first. If the coin falls heads at the first toss, A is already ruined. Surprisingly, however, the expected duration, of the game is quite long: on average one of them is ruined only after 999 trials. (Is this duration really considerably longer than we would expect? In general it can be proved that if A has a dollars, and his adversary B has b, then the average duration of the game is ab trials, especially if $a=b$, then the expected duration of the game is a^2.) *F. Stern* examined the case where the coin is not necessarily true, and called attention in 1975 to a surprising phenomenon (*Math. Mag.* **48**, 286—288.) Suppose A wins with probability p in each turn and B with probability $1-p$ $(0<p<1)$, and they both have a dollars at the beginning of the game. It seems evident that if $p\neq 1/2$, the conditional expectation of the duration of the game, given that A is ruined at the end is completely different from the conditional expectation given that B is ruined at the end of the game. It can still be shown that either A's or B's ruin is assumed, the average durations of the games, and also their distributions are equal. The proof is simple: the probability of B's ruin after the $(2k+a)$th trial is given by $p_{2k+a}=c_{k,a}p^{k+a}(1-p)^k$ $(k=0, 1, 2, \ldots)$, and similarly the probability of A's ruin after the $(2k+a)$th trial is $q_{2k+a}=c_{k,a}p^k(1-p)^{k+a}$ where $c_{k,a}$ is the total number of games which consist of exactly k heads and $k+a$ tails. As the ratio $p_{2k+a}:q_{2k+a}$ is independent of k, the conditional distributions $p_{2k+a}/\sum_k p_{2k+a}$ and $q_{2k+a}/\sum_k q_{2k+a}$ are identical—as we have stated. The explanation of this phenomenon lies in the following fact: if $p=0.99$, then a long game results in B's ruin with great probability, so the expected duration of the game, given that A will be ruined at the end is very short, just as in the case where B's ultimate ruin is given. (For another argument see E. Seneta, "Another look at independence of hitting place and time for simple random walk", *Stoch. Proc. and their Appl.*, **10**, 101—104, (1980).) We have already mentioned that if both A and B have a dollars and they play with a true coin, then the expected duration of the game is a^2 turns; but what happens if they play with two different coins: the probability that A will win with the first coin is $p_1=1/2+\varepsilon$, and with the second coin, it is $p_2=1/2-\varepsilon$, $(0<\varepsilon<$ $<1/2)$. The players choose p_1 or p_2 in each turn, depending randomly on A's accumulated gain k, $(k=1, 2, \ldots, 2a-1)$, in the following way: before starting the game, we draw p_1 or p_2 for each value of k, independently of each other and with equal probability. One may feel intuitively that this game with its complicated formulation is quasi-identical—at least for large a—with the game where $p_1=p_2=1/2$ for every k (i.e., the classical coin tossing game), since the large number of terms $\pm\varepsilon$ equalize each other for large a. But this is not so. *J. G. Sinai* has recently pointed out that the average duration of this complicated game is far longer. Even the logarithm of the average number of necessary tosses is of order $\sqrt{a}$ (in contrast to the above mentioned a^2). This surprising fact can be explained on the basis of Remark (*i*) in I/9. In a sequence of length a, which consist of independent and equally probable p_1's and p_2's, there is a p_1 or p_2 run of length $\log_2 a$ with large probability, and this drifts the gain towards the initial capital, so it delays the ultimate ruin. It is very difficult to get over this "thick wall", and that is why the average duration of the game increases. Problems of this type (i.e., random walks in random environments) are in close connection with the theory of random fields in the last quickie.

e) A paradox of optimal stoppings

We are playing heads or tails with a fair coin so that we stop playing after the nth toss. In this case we win either

$$\frac{n}{n+1}2^n$$

dollars or nothing, depending on whether the outcome is always tails or not. When is it advisable to stop? Let I_n denote our prize (depending on chance) after the nth game:

$$I_n = \frac{n}{n+1}2^n \quad \text{or} \quad I_n = 0.$$

Supposing that $I_n \neq 0$ the expected value of the prize I_{n+1} is

$$E(I_{n+1}|I_n \neq 0) = \frac{n+1}{n+2}2^{n+1}\frac{1}{2}$$

which is greater than $\frac{n}{n+1}2^n$ meaning that it is always worth going on playing. The probability, however, of $I_n = 0$ for some (possibly large) n is 1. Is it really worth playing till we lose everything?

(Ref.: Chow, Y. S., Robbins, H. and Siegmund, D., *Great Expectations: The Theory of Optimal Stopping*, Houghton Mifflin, Boston, 1971.
Shiryayev, A. N., *Optimal Stopping Rules*, Springer, New York, 1978.)

f) The paradox of choices

One often should choose the best one (from a certain point of view) out of a collection of persons or objects (e.g., when shopping or getting married). When studying this problem, we assume that the persons or objects can be arranged in order of goodness, i.e., comparing any two of them, we can always decide which is the better one. Selecting the best would cause no problem if we saw all of them together. In most cases, however, objects or persons have been tried successively and once rejected, one cannot return to that. In the following we will assume that if a "candidate" is not selected when it is his turn then we will not have the opportunity to change our minds later. The problem is not unique even so. We might not even know the total number of opportunities we must choose from. (Generally, we do not have this information when choosing our future wife or husband.) Let us suppose that there are altogether n possibilities, more precisely let n persons or objects pass us in any order (these orders are considered equally probable). Now the question is the following. What method should be chosen to select the best candidate if any of them can only be compared, naturally, with the previous ones. If we always choose, e.g., the third one, the chance of selecting the best is $1/n$. With n growing, $1/n$ converges to 0, and therefore if the number of offers is great, the probability of selecting the best one is nearly 0. Surprisingly, however, there is a method which enables us to select the best candidate with a probability of nearly 30% even if n is a large number. The method is the following. Let the first 37% (more precisely, $100/e$%) of the candidates go and then select the first one better than any previous candidate (if none are better, select the last). In this case the chance of selecting the best is approximately $1/e$, i.e., $\approx 37\%$ however great n is.

If two, three, ..., or generally k choices are allowed and the point is only to have the best one among the k candidates selected, then the optimal probability p_k of this event can be calculated as follows. Let the numbers c_j satisfy the indentity

$$\sum_{j=1}^{\infty} x^{j-1} e^{-c_j x} \equiv 1;$$

then

$$p_k = \sum_{j=1}^{k} e^{-c_j},$$

e.g.,

$$p_2 = \frac{1}{e} + \frac{1}{e^{3/2}},$$

which is more than 1/2(!). It can also be shown that

$$\left(1 - \frac{1}{e}\right)^k \leqq 1 - p_k \leqq e^{-k/e},$$

thus p_k converges to 1 as k tends to infinity.

If the number of candidates N is a random variable, then the chance of selecting the best candidate may decrease. Suppose that the distribution of N_m/m converges to the distribution of a random variable X. Then the optimal probability of selecting the best candidate (more precisely its limit as $m \to \infty$) is

$$p_X = \max_x E(f(x/X))$$

where $f(x) = \max(0, x \ln x)$. The probability p_X may be very small since $\inf_X p_X = 0$.

(Ref.: Chow, Y., Robbins, H., Siegmund, D., *Great Expectations: The Theory of Optimal Stopping*, Houghton Mifflin Co., Boston, 1971. Freeman, P. R., "The secretary problem and its extensions: A review", *Internat. Statist. Review*, **51** 189—206, (1983), Berezovskiĭ, B. A., Gnedin, A. V., *Optimal Choice Problems* (in Russian), Nauka, Moscow, 1984.)

g) The Pinsker paradox of stationary processes

A series of random variables X_n ($n = \dots, -3, -2, -1, 0, 1, 2, 3, \dots$) is called stationary (more precisely, stationary in a wide sense) if firstly, the expected value of X_n does not depend on n (therefore we can assume without the loss of generality that this common expectation is 0) and secondly, the covariances $E(X_n X_m) = r_{n-m}$ (the existence of which is assumed) depend only on the difference $n-m$ (specially if $n = m$, the variances do not depend on n). A vector valued X_n ($n = \dots, -2, -1, 0, 1, 2, \dots$) is stationary if $E(X_n)$ is identically equal to the zero vector and the expected value of the product of the ith coordinate of X_n and the jth coordinate of X_m depend only on i, j and $n-m$. Two basic types of stationary processes are the singular and regular processes. The former is deterministic (i.e., for any value of n, X_{n+1} does not contain any "information" uncorrelated with the random variables preceding X_{n+1}), while the regular type does not have a deterministic part (i.e., if we omit X_n, X_{n-1}, X_{n-2}, etc., then we gradually lose all information). In this way the world of singular processes is ready, and it does not gain information as time passes, while regular processes create a new world out of nothing, that is, the far future is almost independent of the present. (In the Hilbert space of square integrable random variables, i.e., which are of finite variance, the above statement can be formulated as follows. If H_n denotes the subspace which is generated by the random variables preceding X_n then in the singular case $H_n = H_{n-1}$ for all n, while in the regular case $\bigcap_n H_n = 0$.) The importance of singular and regular processes was shown by Wold's theorem, which states that any stationary process can be uniquely decomposed into the sum of a regular and a singular process. It is rather obvious that if X_n is singular then X_{-n} is singular, too, and if X_n is regular then X_{-n} is regular as well. In other words, if n denotes the time parameter then both singularity and regularity remain unchanged when reflecting past and future. Surprisingly, however, it is only true when X_n is a scalar. Pinsker constructed a two-dimensional stationary process which is regular but its inverse (when $-n$ takes the part of n) is already singular. Thus singularity may turn into regularity and *vice versa* if past and future are reversed.

h) The paradox of voting and electing; Random fields

When voting or electing, the outcome is generally uncertain and therefore it is not surprising that important probabilistic results have been discovered in this area too. In 1878, W. A. Whitworth proved the following famous ballot-theorem. If there are two candidates, say, A and B, A scores n votes, B scores m votes, and $n > m$ (i.e. A has won) and p denotes the probability that throughout the counting there are always more votes for A than for B (provided that each order of counting is equally probable), then

$$p = \frac{n-m}{n+m}.$$

Thus, if $n = 2m$, then $p = 1/3$, that is, if A has received twice as many votes as B then the probability that B had an equal number of votes sometime during the counting is twice as much as the probability that A was superior throughout the counting. (See Feller, W., *Probability Theory and Its Applications,* (2nd ed.), Wiley, New York, 1965, p. 66.) This may sound strange but it is not a paradox. Paradoxes do, however, occur in this field too. Marquis de Condorcet (one of Voltaire's friends) pointed out the following example in 1758. (Essai sur l'application de l'analyse à la probabilité des décisions rendues à la pluralité des voix.) If there are three candidates in an election, A, B, and C, and they receive 23, 19, and 18 votes, respectively, then on the basis of majority alone A would be the winner but in actual fact, all the 19 who voted for B may well prefer C to A. In 1950 Kenneth Arrow (the 1972 Nobel Prize winner in economics) used the above example to show that it is logically impossible to create an absolutely fair election system. Thus it is not surprising that there does not exist a standard election system accepted all over the world. (On the probabilistic contraversion of the election system in the USA see Grofman's paper.) The following paradox concerns a special kind of voting: trials. Let us suppose that A, B, C, D and E are the five members of a jury. They decide whether a prisoner is quilty or not by majority. There is a 5% chance that A or B give the wrong verdict, for C and D it is 10%, and E is mistaken with a probability of 20%. (Mistakes are independently committed.) In this case the probability of bringing the wrong verdict is about 0.7%. Paradoxically, this probability increases to about 1.15% if E (who is most probably mistaken) abandons his own judgement and always votes the same way a A (who is most rarely mistaken). The following paradox also shows how surprising situations may arise if voters abandon their own judgements. Let us suppose that each vertex of a planar square lattice is occupied by people who can vote for or against independently of each other with probability p and $1-p$, respectively. Meanwhile each of them chooses one of his four neighbours and votes the next time as that person did previously. The third, fourth, etc. vote is carried out similarly. (When voting for the nth time, everybody gives the $(n-1)$th vote of the chosen neighbour.) The question is the following: What happens if $n \to \infty$? It can be shown that everybody will give the same vote in the end, in other words, "perfect harmony" will be reached. (The probability that everybody votes for or against is p and $1-p$, respectively.) It is worth mentioning that if voters are placed at the vertices of the three-dimensional cubic lattice (where everybody has six neighbours) then such an extreme situation will not occur, i.e., different opinions may harmonize with each other, too (more precisely there is an ergodic limit distribution). The same stands for more than three dimensions. This fundamental difference between two and three dimension is in close connection with the fact (see III/5a) that in the case of two dimensional square lattice, symmetric random walks reach any vertex with probability 1; while in 3 dimensions this is not true. (See Bramson, M., Griffeath, D., "Renormalizing the 3-dimensional voter model", *Annals of Prob.*, 418—432, (1972).)

The above mathematical model of voters standing in the vertices of a square and a cubic lattice, has gained a very important role in the mathematical physics of the last few years. Voters are replaced by "units" with two possible values (e.g., the spin of ferromagnetic materials). These *random fields* are generalizations of stochastic processes in which the time parameter t is replaced by an element of a multidimensional space, e.g., if t stands for the vertices of a d-dimensional cubic lattice and $X(t)$ is a random variable for any t (in the voting model $X(t)$ takes only two values) then $X(t)$ is a random field. Just as we supposed that the voters' opinions are only influenced by those of their neighbours, in physics we may also assume (as a first approach) that each particle is influenced only by its neighbours. This kind of random field is called Markov field (it is the equivalent of Markov chain). In studying ferromagnetism, a special Markov field, namely the Ising model became very important mainly due to the studies of the Norwegian physicochemist *N. Onsager* in 1944. In the last few years Markov fields and especially the Ising model have been applied to help solve the problem of phase transitions. Though the exact notion of Markov field was only introduced in 1968 by the Soviet mathematician *R. L. Dobrushin,* the first description of the notion of phase and that of certain random fields had already been carried out much earlier with potential functions in *J. W. Gibbs*' book in 1902 (Elementary Principles of Statistical Mechanics, Yale Univ. Press). The description of Markov fields by potential functions is especially important because phase transitions occur just when the potential does not determine uniquely the Markov field. In physical terms this means that there may be more than one phase present at the same temperature. The theory also explains why phase transitions are impossible over the critical temperature (even Onsager succeeded in determining the critical temperature). It is interesting that while there cannot occur a phase transition in the one-dimensional model, in the case of the two-dimensional one (on the square lattice) it is already possible. In the later case, in spite of the symmetry of the potential function (the value of the function does not change if all "yes" states are replaced by "no" and *vice versa*), the Markov field itself is not symmetrical. It is due to this paradox (called symmetry-break) that ferromagnetic materials do not lose their magnetism below the critical temperature.

(Ref.: Grofman, B., "Fair appointment and the Banzhaf index", *The Amer. Math. Monthly*, **88**, 1—5, (1981).
Kindermann, R., Snell, J. L., *Markov Random Fields*, Contemporary Math. Vol. 1, AMS, Providence RI, 1980.
Preston, C. J., *Gibbs States on Countable Sets*, Cambridge Univ. Press, 1974.
Sinai, J. G., *Rigorous Results in the Theory of Phase Transitions*, Akadémiai Kiadó, Budapest, 1982.)

Chapter 4

Paradoxes in the foundations of probability theory. Miscellaneous paradoxes

"De naturâ Rationis non est res, ut contingentes; sed, ut necessarias, contemplari."

(B. Spinoza, *Ethica,* Pars Secunda, Propositio XLIV)

"Probability is the most important concept in modern science, especially as nobody has the slightest notion what it means."

(Bertrand Russel, *In a lecture,* 1929)

"Calcul des Probabilités. Première Leçon. 1. L'on ne peut guère donner une définition satisfaisante de la Probabilité..."

(H. Poincaré, *Calcul des Probabilités,* 1896, p. 1.)

"My thesis, paradoxically, and a little provocatively, but nonetheless genuinely, is simply this: PROBABILITY DOES NOT EXIST."

(B. de Finetti, *Theory of Probability,* 1974)

In 1900, at the International Mathematical Congress in Paris, *David Hilbert* considered the problem of the foundation of probability theory as one of the 23 most important unsolved problems in mathematics. Though by the turn of the century probability theory had produced many outstanding results, due to the lack of foundation, this theory as a whole could not join other branches of mathematics. This may be the main cause why *F. Klein,* a professor at University in Göttingen, did not even mention probability theory in his work "Mathematics of the 19th century". Utilizing the results of a number of mathematicians, especially those of *E. Borel, A. Lomnitzky, H. Steinhaus,* and using set and measure theory, *A. N. Kolmogorov* developed the exact theory of probability in 1933. (Details can be found in *Archive for Hist. of Exact Sci.,* **18**, 123—190, 1978.) The base of Kolmogorov's theory is that every event (whose probability we want to obtain—these events are called observable events) can be represented by a subset of the set of all elementary events (i.e., by a subset of the *phase space*). For instance, when tossing a dice, the outcomes can be 1, 2, ..., 6; these elementary events together form the phase space, and the event that the outcome is even can be represented by the subset of the phase space, consisting of the even numbers, {2, 4, 6}. The certain event is represented by the entire phase space, which is traditionally denoted by Ω. Kolmogorov's theory assumes that the observable events form a sigma-algebra (sigma refers to infinity), i.e., the joint occurrence of any two observable events, the occurrence of at least one of finite or countably infinite observable events, and the complement of any observable event is also an observable event. A nonnegative number is assigned to each observable event, this is the probability of the event, such that the probability of the certain event (i.e., of the entire phase space) is 1, and the sigma-additive property holds, i.e., in the case of pairwise exclusive events, the probability of the occurrence of at least one (and so, owing to the pairwise exclusion, exactly one) observable event in a collection of finite or countably infinite observable events is the same as the sum of the probabilities of the observable events in the collection.

The question arises: When defining probability, why do we need sigma-algebras instead of the set of all the subsets of phase space Ω? The answer is very simple: In general, the probability cannot be defined on the set of all the subsets of Ω, more precisely, if probability is defined on a sigma-algebra consisting of some subsets of Ω, then this probability may not be extended to the rest of the subsets of Ω if sigma-additivity is still required (unless Ω consists of finite or countably infinite elements). *G. Vitali* knew this result as early as 1905. Let the phase space be the interval (0, 1), and make an attempt to define the probability on all the subsets of (0, 1) according to the "uniform distribution". Obviously, the probability $b-a$ should be assigned to a subinterval (a, b) Thus, due to sigma-additivity, the probability is automatically defined on the least sigma-algebra containing the intervals. This probability can be extended to some other sets, but there also exist sets to which it cannot be extended, i.e., on which the probability cannot be defined according to the "uniform distribution".

Such a "pathological" set was constructed by *E. Zermelo* as follows. He selected the points of the interval (0, 1) into disjoint classes such that the points whose distance was rational belonged to the same class. Then, using the axiom of choice, he defined a set H that had exactly one point from each of the above classes. It can be proved that this set H cannot have any probability according to the "uniform distribution".

It can also be shown that if we abandon "uniformness" but require that each subset of Ω have probability and the probability of each point in Ω be 0, then even this kind of probability definition is impossible in the case of a phase space Ω whose cardinality is countable or—assuming the continuum hypothesis—continuum (see G. Birkhoff, *Lattice Theory,* Amer. Math. Soc., Providence, 1967, p. 266). It is not yet known whether or not a space Ω (with sufficiently large cardinality) exist such that there can be defined a probability fulfilling the above requirement. This is the *problem of measurable cardinalities.* The situation changes crucially if we abandon the axiom of choice; see T. Jech; *Set Theory,* Acad. Press, New York, 1978 and R. M. Solovay; "A model of set theory in which every set of reals in Lebesgue measurable", *Annals of Math.,* 1—56, (1970).

Though in Kolmogorov's theory the probability is always a nonnegative number, several theorems in probability theory can be extended so that a negative number can be a probability, too. For example *K. J. Hochberg* (*Proc. Amer. Math. Soc.,* **79**, 298—302, 1980) proved that in the theorems obtained by such an extension of the central limit theorems, there occurs the real (both positive and negative) valued "density function" $u_n(t, x)$ which can be derived from the fundamental solutions to the following extension of the differential equation of heat conducting

$$\frac{\partial u}{\partial t} = (-1)^{n+1} \frac{\partial^{2n} u}{\partial x^{2n}}$$

$n=2, 3, \ldots$ (the original differential equation of heat conducting is the case where $n=1$). This book, however, does not deal with negative nor complex valued probability measures, neither discusses other extensions of probability.

1. PARADOXES OF RANDOM NATURAL NUMBERS

a) The history of the paradox

In Kolmogorov's theory of probability it is impossible to choose a natural (positive integer) number at random with uniform distribution, for if the probability of selecting, e.g., 1 is 0, then due to uniformity, the probability of choosing any other natural number is also 0. Thus sigma-additivity leads to a contradiction because the probability of choosing a natural number is 1 and not 0. On the other hand, if the probability of choosing 1 is positive then sigma-additivity leads again to a contradiction (the probability of the certain event would be infinite). In spite of this fact it is natural to expect that the probability of choosing an odd or an even number is 1/2. The next definition (which disregards of sigma-additivity) gives just this probability. Let K be an arbitrary subset of natural numbers and let k_n denote the number of elements in K not greater than n. The relative frequency k_n/n shows the probability of choosing a number from K provided we may choose with uniform distribution from the first n numbers. If the limit of the relative frequency k_n/n when n tends to infinity exists then this limit is called the probability of K. By this definition the probability of choosing an integer divisible by 2, 3, etc. is 1/2, 1/3, ..., respectively. The probability that two random integers (chosen independently with uniform distribution) are relative primes can also easily be calculated. Supposing firstly that none of the integers is greater than n, the corresponding probability (depending on n) is calculated, and then its limit is considered as $n \to \infty$. Čebyshev already showed in the last century that this limit is $6/\pi^2$ ($\approx 2/3$). Accordingly, if both the numerator and the denominator of a fraction are random natural numbers then it cannot be reduced with probability $6/\pi^2$. The following paradoxes also concern random natural numbers. According to *J. E. Littlewood,* the first is due to the famous physicist, *E. Schrödinger*. In a 1935 article *F. P. Cantelli* attributed the second paradox to *P. Lévy*. (P. Lévy was one of the most outstanding geniuses of probability theory. He came to occupy Poincaré's and Hadamard's seat at the French Académie des Sciences.)

b) The paradoxes

(*i*) One of two consecutive random natural numbers are drawn on the foreheads of two players A and B. The person with the smaller number loses and is obliged to pay the other as many dollars as shown on his own forehead. Both players have the right to veto, i.e., finding the number on the other's forehead too large, any of them can ask for a new game. (The number drawn on his own forehead is naturally unknown to the player.) However, following the reasoning below, none of them will veto. Both of them may think: "I can see the number k on my opponent's forehead. Therefore I have either $k-1$ or $k+1$. Each case is equally probable but if I lose, I pay only $k-1$ dollars, while if I win, I get k dollars, so it is not worth vetoing." As the expected value of the prize is positive the game seems to be favourable to both players, which is of course impossible.

(*ii*) Let us choose two random natural numbers X and Y independently, with uniform distribution. For any fixed (non-random) number x the probability of $Y \leqq x$ is 0. Similarly, for any fixed y the probability of $X \leqq y$ is 0. Consequently, the probability of both $Y \leqq X$ and $X \leqq Y$ are also 0, which is impossible, for one of them is certainly true.

c) The explanation of the paradoxes

(*i*) The paradox is brought about the fact that there is no uniform distribution on the set of natural numbers. If the numbers written on the player's foreheads were at most 3-digit-numbers then there would already exist a uniform distribution on these numbers, but then the above reasoning that led to the paradox would become completely false.

(*ii*) No doubt, the probability of $Y \leqq x$ is 0 for any fixed x (by the definition mentioned in the history of the paradoxes), but from this fact it does not follow that the probability of $Y \leqq X$ is also 0. It would follow only if the probability were sigma-additive, but this kind of probability (as we have mentioned) is not sigma-additive.

d) Remarks

(*i*) Number theory and probability theory are in close connection. To illustrate how probabilistic ideas can be applied in number theory, we recall first that the relative frequency of primes among integers less than n is about $1/\log n$ (if n is large enough). Supposing that primes are distributed randomly and independently among the first n numbers, the probability of choosing two primes close to n is about $1/(\log n)^2$ (due to the independence). Let us consider an interval around n the length of which is c. (c is small compared to n, but large enough for statistical considerations.) According to the above result, the number of twin primes (primes the difference of which is 2) belonging to this interval is $c/(\log n)^2$. A more detailed analysis (which takes into account, e.g., that an integer differing by 2 from a prime ($\neq 2$) is certainly odd and therefore more likely a prime itself) shows that the expected number of twin primes exceeds $c/(\log n)^2$ by about 32%. Calculating on this basis, *M. F. Jones, M. Lal* and *W. J. Blundon* published a table in the Mathematics of Computation in 1967 which shows, e.g., that among the first 150 thousand numbers greater than 100 million the expected number of twin primes is 584. Actually this number is 601. The difference is fairly small. Similarly, considering the first 150 thousand numbers after 100 trillion, we expect 191 twin primes, while the actual number is 186. This kind of "statistical" approach to primes produces fairly good results. It is of special interest because the number of twin primes (infinite or not) is still unknown. (The largest prime known up to the present is $2^{86243}-1$). Primes follow each other according to a very complicated seemingly random rule. This is why a probabilistic approach is excellent in this case. We will return to the relation of complexity and randomness with "the paradox of the Monte Carlo method".

(*ii*) According to the finitely additive uniform distribution on the natural numbers, the probability of any finite subset (of natural numbers) A is 0. Supposing now that the distribution is not uniform but it has the property that given an arbitrary positive number ε, there is a finite set A whose probability is $P(A)>1-\varepsilon$, then the difference between additivity and sigma-additivity disappears. More precisely, if the probability P is additive on the finite subsets of natural numbers (or on any countable set

Ω), then the probability can be extended to every subset in such a way that the extension becomes sigma-additive on the sigma-algebra of all subsets of natural numbers.

If Ω is not a countable set (e.g., the whole interval (0, 1)) then quite strange additive probabilities may occur. They may take only the values 0 and 1, and they are defined on every subset of (0, 1) (supposing the axiom of choice in set theory). These probabilities are strange because countably many events with probability 1 may very unlikely to occur simultaneously, i.e., this probability might be 0. Similarly, at least one of countably many events with probability 0 may occur with probability 1.

e) References

Elliot, P., *Probabilistic Number Theory,* Springer, New York, 1980.

Kac, M., "Statistical independence in probability, analysis and number theory", Carus, *Math. Monographs,* **12**, Wiley, New York, (1959).

Rényi, A., "On a new axiomatic theory of probability", *Acta Math. Acad. Sci. Hung.,* **6**, 285—335, (1959).

The following paper is a paradoxical approach of an outstanding unsolved problem. It claims that the *Riemann* hypothesis (which does not depend on chance!) is true with probability one:

Good, I. J., Churchhouse, R. F., "The Riemann hypothesis and pseudorandom features of the Möbius sequence", *Mathematics of Computation,* **22**, 857—864, (1968).

A random analogue of Dirichlet's celebrated theorem on the infinitude of primes in arithmetic progressions is discussed in:

Ruzsa, I. Z., Székely, G. J., "Intersections of traces of random walks with fixed sets", *Annals of Probability,* **10**, 132—136, (1982).

2. BANACH—TARSKI PARADOX

a) The history of the paradox

The uniform probability, or the corresponding length, area and volume in one, two, and three dimensions cannot be defined on arbitrary sets if the *sigma-additivity* of these measures is required. However, if we assume only *additivity,* (that is, the measure of the union of two disjoint sets equals the sum of their individual measures), then—as the Polish mathematician *S. Banach* showed—in one and two dimensions every bounded set becomes measurable (has length, or area). Thus, accordingly, uniform probability can also be defined on every (bounded) set in one and two dimensions if we assume only the additivity of probabilities. *Hausdorff*, however, showed in 1914 that such extension of measures in three dimensions is impossible. *S. Banach* and *A. Tarski* set forth a paradoxical theorem in 1924 which picturesquely showed that neither an additive measure (volume), nor the corresponding uniform probability can be defined on arbitrary bounded sets in three dimensions.

b) The paradox

Considering a ball of radius $r=1$ cm, it is possible to divide it into some finite number of pieces and then reassemble them to form a ball of radius $R=1$ km. In general, if A and B are bounded subsets of R^3 having nonempty interiors, then there exist a natural number n and partitions $\{A_j:1\leqq j\leqq n\}$ and $\{B_j:1\leqq j\leqq n\}$ of A and B, respectively, into n pieces each, such that A_j is congruent to B_j for all j. (A subset X of R^3 is bounded if it is contained in some ball, and X has a nonempty interior, if it contains some ball. By a partition of a set X we mean a pairwise disjoint family of subsets of X whose union is X.)

c) The explanation of the paradox

If we chop a ball of radius $r=1$ cm into a finite number of pieces, we might intuitively expect that putting the pieces together, they can only form solid figures whose volume is equal to that of the original ball of radius 1 cm. This is, however, true only if we chop the ball into pieces which have volume. The point of the paradox is that in the three-dimensional space there are non-measurable sets, to which we cannot assign volume, if we want to keep the additive property of the volume, and if we want the volumes of two congruent sets to be equal. (The proof of the Banach—Tarski theorem depends on Zermelo's Axiom of Choice.)

d) Remarks

Several outstanding mathematicians (for example, the Italian *de Finetti*) consider the sigma-additivity of probability a too strong restriction, but accept additivity. The Banach—Tarski paradox shows that changing sigma-additivity for additivity does not solve every problem and also brings new ones. In automata theory the dilemma of assuming or not assuming sigma-additivity became so critical, that even the Encyclopaedia Britannica deals with the problem. Electronic computers are often used to generate (theoretically) infinite sequences of random numbers (cf. the next paradox). The probability of each sequence is zero, but the probability of their union is one. Thus the acceptance of sigma-additivity rests upon the tacit assumption that we cannot generate random phenomena by automata, i.e., random and non-random sequences are separated, which is just the well-known bifurcation of the Greek goddesses Tyche (chance) and Moira (fate).

e) References

Banach, S., Tarski, A. "Sur la décomposition des ensembles de points en parties respectivement congruentes", *Fund. Math.*, **6**, 244—277, (1924).

Stromberg, K., "The Banach—Tarski paradox", *The American Math. Monthly*, **86**, 151—160, (1979).

3. THE PARADOX OF THE MONTE CARLO METHOD

a) The history of the paradox

The Monte Carlo Method is a numerical method based on random sampling. In solving numerical problems there can frequently be found a probabilistic model where the unknown number appears. Then it is possible to solve the problem in such a way that we observe the outcomes of random experiments belonging to the probabilistic model so many times that we can estimate (from these outcomes) the unknown number with a prescribed accuracy. Though the idea of this method is quite old, its actual application dates back only to the invention of computers when

J. Neumann, S. Ulam and *E. Fermi* used it for the approximate solution of difficult numerical problems of nuclear reactions. The name of the method refers to the series of random numbers used here which, in principle, could also be the regularly announced results of games played in gambling houses, e.g., in Monte Carlo. In practice, however, the computer itself produces the random numbers necessary for the method. Consequently this nice name (first used by *N. Metropolis* and *S. Ulam* in 1949) is totally misleading (the method is not of much help in trying to win in Monte Carlo). The idea of Monte Carlo method first appeared in a 1777 work by *Buffon* (see I. 11). It gives a method for the estimation of π by throwing a needle randomly. If parallels are drawn on a table at unit distance and a needle of length $L<1$ is thrown randomly on the table (the angle between the parallels and the needle, and the distance of the centre of the needle from any given parallel are independent and uniformly distributed over $(0,2\pi)$ and $(-1/2, 1/2)$, respectively) then the probability that the needle will intersect one of the parallels is $2L/\pi$. If the experiment is carried out many times then the relative frequency of intersections will be very near to the theoretical probability $2L/\pi$, and thus π can be calculated. This method of the approximation of π is only of theoretical importance since to get two-figure accuracy, several thousands of throws have to be made. (By another method π can be determined to one million figures, see *G. Miel*'s article.) Buffon's needle problem shows that the Monte Carlo method is not suitable for very accurate calculations. Even to obtain results of two or three-figure accuracy, thousands or millions of experiments have to be made. It is obvious therefore that the Monte Carlo method only became applicable when experiments could be simulated by computers. Instead of needle-throwing, two independent random numbers were generated which determined the position of the (supposed) needle and whether it intersected the (supposed) parallels. As the computer is able to generate several millions of numbers a minute, it does not take too long to simulate millions of experiments that would otherwise take a life-time.

The theory of generating random numbers by computers has become an important branch of mathematics. Instead of actual random numbers (which might be produced by any random physical process such as radioactive decay), pseudo-random numbers, generated by deterministic computer algorithms, came into the limelight.

In connection with pseudo-random numbers the following question arises. In what sense can they be considered random since they are generated by deterministic (non-random) algorithms? Since *von Mises*' article in 1919 several outstanding mathematicians have dealt with this problem. (Its philosophical aspects were studied by *P. Kirschenmann* and *P. McShane,* among others.)

b) The paradox

In 1965—66 *Kolmogorov* and *Martin-Löf* put the notion of randomness in a new light. They defined when a series consisting of 0's and 1's can be considered random. The main idea is the following. The more difficult it is to describe a series (i.e., the longer its "shortest" generating program is) the more random we may consider it. Naturally, the length of this "shortest" program may vary if we use different computers. For this reason a standard machine is chosen which is called Turing-machine. The measure of complexity of a series is the length of the shortest Turing program which can generate the series. Complexity is a measure of irregularity. A series whose length is N is called random if its complexity is nearly maximal. (It can be shown that most series are of that kind.) As Martin-Löf proved, these series can be considered random because they satisfy all the statistical tests of randomness. Complexity and randomness are therefore in close connection. If a programmer wants to generate "real" random numbers, then, due to Kolmogorov's and Martin-Löf's results, he can only generate the series by a rather long program. At the same time, in practice, random number generators are very short. How can these two things be reconciled?

c) The explanation of the paradox

Series generated by short programs and used as random numbers actually satisfy only a few criteria of randomness, not all. This, however, causes hardly any problems in application. For example, for the purpose of numerical integration, it is enough to know that the pseudo-random numbers are uniformly distributed over an interval.

Suppose we want to integrate a function of bounded variation on the interval (0, 1). Then the number

$$I = \int_0^1 f(x)\,dx$$

is approximated by the mean

$$I_N = \frac{1}{N} \sum_{i=1}^{N} f(x_i)$$

not only if the series $x_1, x_2, \ldots, x_N$ is random and uniformly distributed over interval (0.1). It is enough to require that the series is uniformly distributed. This means that as $N \to \infty$,

$$D_N = \sup_{0 \leq x \leq 1} |c(x, N) - x|$$

converges to 0, where $c(x, N)$ is the quotient of the number of $x_1, x_2, \ldots$ $\ldots, x_N$ belonging to $(0, x)$ and N, i.e., the relative frequency.

It can be shown that

$$|I - I_N| \leqq V_f D_N$$

where V_f is a constant depending on function f (the total variation of f). From this it follows that the approximation of I is the more accurate the less D_N is. D_N, however, is not minimal in the case of random series. For random series the order of approximation is $N^{-1/2}$, while in non-random cases an accuracy of $N^{-1} \log N$ can be obtained.

In many cases it turns out that instead of trying to cope with the "impalpable concept of randomness" one should use deterministic sequences that are very well suited for given problems. This is the essence of the quasi-Monte-Carlo-method.

d) Remarks

(*i*) Recently connections between randomness and complexity have led to several interesting discoveries. In mathematics it has long been a general practice to handle too complicated structures as if they were random (e.g., the behaviour of the complicated sequence of primes is frequently described by probabilistic laws). The concept that randomness cannot be distinguished from complexity is, however, such a revolutionary idea that it is significant even from a philosophical point of view. Using this concept, Spinoza's motto can be restated as follows: People prefer simple things to complicated ones—which is undoubtedly true. At the same time it is obvious that the more we try to understand nature the more we have to realize that not everything is simple.

(*ii*) The application of random number series is rather wide-ranging. Numerical integration, numerical solution of differential equations, computer simulation of physical, chemical, biological, technical, and economic problems, etc. also require random sequences. They help to solve traffic, transport, and other optimalization problems, as well as creating astronomical models. The efficiency of different computer programs can also be tested if the data are random numbers.

Finally we should mention a completely different field of application, the computer art, where random number sequences offer millions of variations (random number sequences can of course correspond to series of sounds, colours, letters, etc.). From random sequences the computer filters out those that do not meet the rules recognized when studying sample models. If the computer works on the base of enough samples, the artistic result will be fairly good. Xenakis, the Greek composer has used, e.g., computer-made random sounds in his works. Several exhibitions have already been organized from computer graphics. (It has to be noted that not all computer graphics apply random sequences.) An international organization of computer artists was founded in 1970.

e) References

Kirschenmann, P., "Concept of randomness", *J. Phil. Logic,* **1**, 395—414, (1972).

Knuth, D. E., The Art of Computer Programming, (Chapter 3—Random numbers), Addison-Wesley, 1969.

Martin-Löf, M., "The definition of random sequences", *Information and Control,* **9**, 602—619, (1966).

McShane, P., *Randomness, Statistics and Emergence,* Univ. Notre Dame, 1970.

Metropolis, N., S. M. Ulam, "The Monte Carlo method", *J. Amer. Statist. Assoc.*, **44**, 335—341, (1949).
Miel, G., "An algorithm for calculation of π", *The American Math. Monthly*, **86**, (1979).
Niederreiter, H., "Quasi-Monte Carlo methods and pseudorandom numbers", *Bull. Amer. Math. Soc.*, **84**, 957—1041, (1978).
Schnorr, C. P., *Zufälligkeit und Wahrscheinlichkeit.* Lecture Notes in Math., **218**, Springer, Berlin—New York, 1971.
Sobol, I. M., *The Monte Carlo Method*, Mir, Moscow, 1975 (in Russian).
Székely, G. J., Tusnády, G., "On the philosophical concept of randomness from a mathematical point of view", (in Hungarian), *preprint*.

4. THE PARADOX OF UNINTERESTING NUMBERS; AN INCALCULABLE PROBABILITY

a) The history of the paradox

Whether a number is interesting or uninteresting is completely subjective, but one can give an objective definition on the basis of the previous paradox. We shall consider a number interesting if its complexity (defined in the previous section) is small. Therefore rational numbers are interesting, for their decimals recur periodically; π and e are also interesting among irrational numbers, since their digits can be generated by a quite simple computer program. There exist however irrational numbers which are more irregular. Normal numbers, for example, have the following property: every decimal (and what is more, every group of fixed number of digits) occur with the same probability in the infinite sequence of their decimal expansion. Most of the irrational numbers are normal, but it is difficult to decide whether a particular number is normal or not. Thus, for example, it is not known whether π (whose first one million decimals were published 1974) or e are normal or not. At the same time there exists a very simple (but artificially constructed) example of a normal number. In the early thirties *D. G. Champernowne* showed that the following number is normal:

$$0.123\,456\,789\,101\,112\,131\,415\,161\,718\,192\,021\,222\,324\,252\,6\ldots$$

(the decimals are the consecutive natural numbers). The situation was similar in arithmetic more than hundred years ago, when *Liouville* constructed a transcendental number for the first in 1844, that is, a number which cannot be the solution of an algebraic equation with integral coefficients. π and e were only proved to be transcendental numbers as late as 1882 and 1873 by *F. Lindemann* and *Ch. Hermite.* The study of numbers regarding normality began only after the turn of the century, due especially to the researches of *E. Borel.* Since that time the investigation of regularity and irregularity in the sequences of digits evolved into an interesting theory, especially after the researches of *A. N. Kolmogorov, P. Marin-Löf, R. J. Solomonoff* and *G. J. Chaitin.* The following paradox is one among the many paradoxes in this field.

b) The paradox

In most numbers digits follow each other randomly, that is, most of the numbers are uninteresting in the following sense: the computer programs which produce these numbers are not much shorter than the numbers themselves. In spite of this most numbers cannot be proved to be uninteresting (in any system of axioms free from contradiction). There exist an infinite number of uninteresting numbers, but this can be proved only for a finite number of them.

c) The explanation of the paradox

Initially it may seem surprising that something that cannot be identified may exist, but similar phenomena appear not only in the world of mathematics. For example, if all the one hundred thousand seats in a stadium are occupied, but only ninety-nine thousand tickets have been sold, then it is clear that one thousand people have sneaked in without a ticket. The identification of these people, however, is hopeless (especially if the tickets were taken away from everybody at the entrance). Thus we are sure that there are one thousand people in the stadium who got in without a ticket, but we cannot prove that any particular person got in without a ticket. Phenomena such as this occur frequently in mathematics.

It is not at all surprising that most of the numbers are uninteresting, if we consider the fact how difficult it is to "discover" any regularity even in a seven-digit telephone number, to make it easier to memorize. In the case of one hundred, or thousand digit numbers this would be even more difficult for a larger percent of these numbers. So it is the second part of the paradox which is more surprising, especially, for those who are inexperienced in the paradoxes of 20th century logic. Among these paradoxes *G. G. Berry's* is the nearest to ours. (This paradox was published for the first time seventy years ago in "Principia Matematica" by *B. Russel* and *A. N. Whitehead.*) The "computerized version" of Berry's paradox comes from *E. F. Beckenbach.* It claims that uninteresting natural numbers may not exist, because then the smallest of these would be interesting. In other words: the smallest of the numbers which can be produced only by long computer programs can also be produced by a short program, and this is undoubtedly a contradiction. It must be assumed that some numbers can be uninteresting even if we cannot prove it. One can show that if the system of axioms and inference rules we use contains n bits of information, then the "uninteresting" property of a number cannot be proved if its information content is much more than n bits.

d) Remark

A very important criterion of the randomness of digits in a number is that they cannot be extrapolated, or predicted. The question is whether there exist any (not random) number which can be defined precisely but whose digits cannot be predicted. The question was answered in the affirmative in an example of *Chaitin.* Let a random heads-tails sequence, or a corresponding 0—1 sequence be the input of a certain computer, namely a Turing-machine. The probability that the Turing-machine will ever stop for a random input, defines the Chaitin number. (Theoretically the machine may work for an infinitely long time, because it does not receive order which would make it stop.) It can be proved that the Chaitin number is an "uninteresting" number whose decimals cannot be predicted. At the same time this "uninteresting" Chaitin number has very interesting properties. If, for example, we knew its first few thousand decimals, then we would also get the answers to some classical, unsolved problems of mathematics, such as the *Fermat-conjecture* or the *Goldbach-conjecture.* The Fermat conjecture, (which claims that the equation $x^n+y^n=z^n$

cannot be solved for natural numbers x, y, z and $n>2$)*, could be proved or disproved theoretically by a computer program, the "Fermat-program" which would compute for given values of n and z if there exist any numbers x and y which are solutions to the Fermat equation. If we gradually increased the values of n and z, then every case would be checked. The computer would stop if it found a solution. If the computer ever stops, the conjecture is disproved, and if it never stops, the conjecture is true. The "only" problem is the following: no matter how long the computer has been already working for, we can never be sure that the machine will not stop in the next step. We could get round this problem if we knew the Chaitin-constant. Consider all the binary inputs of finite length and try to select the programs (inputs) which terminate the computer. First we try to see if the computer stops for the first program in the first step; then if it stops for the second program in the first step. Then we let the first program run till the second step, the third program till the first step, the second one till the second step and the first one till the third step, etc. If the computer stops for some binary input of length k, then in thought we put a $1/2^k$ unit weight into a sack. Gradually there will be more and more weights in the sack and their sum will converge to the Chaitin-constant (since the probability that an arbitrary binary sequence of length k will occur in a heads-tails sequence is $1/2^k$). Let m be the length of the binary "Fermat-program". We continue to run the programs until the difference between the Chaitin-constant and the accumulated weight in the sack is less than $1/2^m$. If the Fermat-conjecture has not turned out to be false up till this time, it must be true, because if the "Fermat-program" terminated the computer later, then we would have to put a weight of $1/2^m$ units into the sack, and this is in contradiction with the fact that we have approached the Chaitin-constant with an accuracy of more than $1/2^m$. The Chaitin-constant contains the solutions (or the theoretical possibility of solutions) for all the problems which can be reduced to a stopping (halting) problem such as the one we have just discussed.

* Recently a German mathematician, *G. Faltings* has proved a very deep theorem implying that the possible (essentially different) solutions of the Fermat's equation are finite, a big step in Fermat's direction.

e) References

Chaitin, G. J., "Randomness and mathematical proof", *Sci. Amer.*, **232**, 47—52, (1975).
Gardner, M., "The random number omega bids fair to hold the mysteries of the universe," *Sci. Amer.*, **241**, 22—31, (1979).
Guilloud, J., Bouyer, M., *Un million de decimals de π. Paris*, 1974.
Miel, G., "An algorithm for calculation of π", *The Amer. Math. Monthly*, **86**, (8), (1979).
Shanks, D., Wrench, J. W., "Calculation of π to 100 000 decimals", *Mathematics of Computation*, **16**, 76—99, (1962).

5. THE PARADOX OF RANDOM GRAPHS

a) The history of the paradox

Structural problems in several fields of science (such as the problem of electrical network) can easily be demonstrated and solved by graphs, i.e., by points and lines connecting them. Points are called the vertices, lines are called the edges of the graph. Edges may also represent connections depending on chance. That is why the research concerning the structure of random graphs is of great importance. The theory of random graphs is mainly due to the work of *Paul Erdős* and *Alfréd Rényi*.

Suppose that a graph has n vertices and each edge is drawn with probability p independently of the existence of other edges. Let $\varepsilon > 0$ be an arbitrary number. In 1960 Erdős and Rényi proved that if

$$p \leqq \frac{(1-\varepsilon)\log_2 n}{n}$$

then the probability that the graph is connected converges to 0 as n increases; on the other hand, if

$$p \geqq \frac{(1+\varepsilon)\log_2 n}{n}$$

then this probability converges to 1. (We say that a graph is connected if any vertex can be reached from any other vertex through the edges.)

Thus the probability

$$\frac{\log_2 n}{n}$$

has a "dividing ridge" role. During the last two decades, the theory of random graphs was extended to graphs with infinite vertices. In connection with these infinite graphs, Erdős and Rényi drew attention to the following paradox.

b) The paradox

We say that two graphs G_1 and G_2 are isomorphic if there exists a one-to-one correspondence between the vertices of G_1 and those of G_2 such that two vertices in G_1 are connected if and only if the corresponding vertices in G_2 are also connected.

If two graphs are isomorphic then their vertices have the same cardinality, but this is by no means a sufficient condition for isomorphism. However, if the cardinality is infinite, more precisely, if the cardinality of the vertices is the same as that of the integers and any two of them are connected with probability 1/2 independently of the other edges, then these graphs are isomorphic with probability 1. Consequently, in this sense all infinite random graphs are the same!

c) The explanation of the paradox

We say that a graph is universal if for any sequences $u_1, u_2, \ldots, u_n$ and $v_1, v_2, \ldots, v_n$ of vertices (different from each other) there exists a vertex w different from the u's and v's such that w is connected with every u but with none of the v's. It is easy to show that if G_1 and G_2 are universal then they are also isomorphic. The probability that the random graphs in the paradox are not universal, however, is 0 (i.e., w exists with probability 1).

d) Remarks

Beside the research of random graphs, the analysis of other random structures (random matrices, random algebraic equations, random power series, etc.) has also led to several interesting results in the last few years.

E.g., *N. B. Maslova* proved the following theorem: If the coefficients X_j of the random algebraic equation

$$\sum_{j=1}^{n} X_j z^j = 0$$

are independent and identically distributed random variables with expectation 0 (but are not identically 0) and

$$E(|X_j|^{2+\varepsilon}) < \infty$$

for some positive ε, then, asymptotically, the number of its real roots is normally distributed with expected value

$$\frac{2}{\pi} \ln n$$

and standard deviation

$$2\sqrt{\pi^{-1}(1-2\pi^{-1})\ln n}.$$

e) References

Erdős, P., Spencer, J., *Probabilistic Methods in Combinatorics*, Akadémiai Kiadó, Budapest, 1974.

Maslova, N. B., "On the distribution of the number of real roots of random polynomials", (in Russian), *Theory of Prob. and its Appl.*, **19**, 488—500, (1974).

Mehta, M. L., *Random Matrices*, Academic Press, New York, 1967.

6. THE PARADOX OF EXPECTATION

a) The history of the paradox

A well-known theorem of probability theory states that if X and Y are random variables with finite expectations, then the expectation of their sum exists and equals the sum of their expectations: $E(X+Y)=$ $=E(X)+E(Y)$. It can easily be shown that even if $E(X)$ and $E(Y)$ do not exist, but $E(X+Y)$ exists, then $E(X+Y)$ depends only on the distributions of X and Y, that is, $E(X+Y)$ can be determined without knowing the joint distribution of X and Y. Surprisingly this is not true for three variables.

b) The paradox

If X, Y and Z are arbitrary random variables, for which $E(X+Y+Z)$ exists, then this expectation cannot always be determined knowing only the individual distributions of X, Y and Z.

c) The explanation of the paradox

Let us define the random variables X, Y and Z in two different ways. Their distribution will be the same in both cases, but the expectation $E(X+Y+Z)$ will be different.

Let U be uniformly distributed on the interval $(0, 1)$. Then clearly $1-U$ and $V=(2U-1)$ are also uniformly distributed on $(0, 1)$. If

$$X = Y = \operatorname{tg}\left(\frac{\pi}{2} U\right)$$

and $Z=-2X$, then $X+Y+Z\equiv 0$ and thus $E(X+Y+Z)=0$, whereas if

$$X = \operatorname{tg}\left(\frac{\pi}{2} U\right), \quad Y = \operatorname{tg}\left(\frac{\pi}{2}(1-U)\right)$$

and

$$Z = -2 \operatorname{tg}\left(\frac{\pi}{2} V\right),$$

then the inequality $X+Y+Z>0$ holds with probability one, so $E(X+$ $+Y+Z)$ is also positive, more precisely,

$$E(X+Y+Z) = \frac{4}{\pi} \ln 2.$$

d) Remarks

(*i*) Since $E(X+Y+Z)=E((X+Y))+Z$ and $E(X+Y+Z+W)=$ $=E((X+Y)+(Z+W))$, the expectations of sums of three and four variables are uniquely determined by the two-dimensional distributions. It is not known, however, whether this is true for more than four random variables or not.

(*ii*) *Ruzsa* and *Székely* showed that it is possible to assign a real number $E(X)$ to every random variable X, so that this number equal their expectations if they exist, and are finite; and

$$E(X+Y) = E(X)+E(Y)$$

always holds if X and Y are independent. Our paradox shows that this kind of extended expectation does not exist for not necessarily independent random variables [for the random variables defined in Section c) $E(X+Y+Z)$ should be zero since $E(X)=E(Y)$ and $E(Z)=-2E(X)$].

e) *References*

Ruzsa, I. Z., Székely, G. J., "An extension of expectation", *Zeitsch' Wahrsch' theorie verw. Geb.*, **53**, 17—20, (1980).
Simons, G., *"An unexpected expectation"*, *Annals of Prob.*, **5**, 157—158, (1977).

7. THE PARADOX OF THE FIRST DIGIT

a) *The history of the paradox*

About a century ago, in 1881, *Simon Newcomb* drew attention to an interesting empirical fact in the Amer. J. Math. This discovery was soon forgotten, however, and was only rediscovered 60 years later by *Frank Benford,* a physicist at the General Electric Company. The law was named after him. (Newcomb is not the only person to have been unfairly treated. The sarcastic law of Eponymy states that no scientific theorem or discovery is named after its original discoverer.) W. Weaver tells Benford's story in "Lady Luck": "I have been told that an engineer at the General Electric Company, some twenty-five years or so ago, was walking back to his office with a book containing a large table of logarithms. He was holding it at his side, spine down; and as he glanced down at the edges of the pages, he noticed that the book was dirtiest at the opening pages and became progressively cleaner—just as though the early parts of the book had been consulted a lot, the middle less, and the concluding part least of all. 'But that', he must have thought, 'is ridiculous. That implies that people must frequently look up the logarithms of numbers beginning with the digit 1, next most frequently numbers beginning with 2, and so on, and least frequently numbers beginning with 9.

And this just cannot be so; because people look up the logarithms of all sorts of numbers, so that the various digits ought to be equally well represented."

b) The paradox

Consider a table, e.g., the table of the integer powers of 2 or any table of physical constants or a table of population statistics. It generally turns out that the first digit ($\neq 0$) of the numbers in the tables is not uniformly distributed on 1, 2, 3, ..., 9. 1 is the most frequent, then comes 2 and so on, 9 being the rarest. According to Benford, the relative frequency of the first digits not greater than k is not $k/9$ (which would mean uniform distribution) but rather $\lg(k+1)$ (where lg stands for $\log_{10}$). Consequently, the relative frequency of 1, 2, ..., 9 is about 30%, 17%, ..., 5%. (Benford's law can be put in another way, i.e., the mantissas of the logarithm of numbers are of about uniform distribution over the interval (0, 1).) Benford's law does not claim that 1 is the most frequent first digit in every table (anybody could create a table containing not a single 1) but that typically the tables contain more ones as first digit than, e.g., nines.

c) The explanation of the paradox

There are several probabilistic and non-probabilistic approaches to Benford's law. Consider first a non-probabilistic one.

Let us examine the table of the powers of 2. The first digit of 2^n is 1 if there exists and integer s such that $10^s \leqq 2^n < 2 \cdot 10^s$. If n (and therefore s) is large enough then s/n is approximately equal to $\lg 2$, which means that among the first n powers of 2 every $\lg 2$-th begins with 1. Similarly, the rate of powers of 2 beginning at most with k is about $\lg(k+1)$ as in Benford's law.

The probabilistic approach is a bit more complicated. Again we have to start from the fact that the first nonzero digit of a positive random number X is at most k if there exists an integer s such that

$$10^s \leqq X < (k+1) \cdot 10^s.$$

Consequently, Benford's law holds only if the probability of the fractional part $\{\lg X\}$ being at most $\lg(k+1)$ is exactly $\lg(k+1)$. A sufficient condition is that the fractional part $\{\lg X\}$ is uniformly distributed over the interval (0, 1). Now the first question is the following: What conditions on the distribution of X imply that the distribution of $\{\lg X\}$ is approximately uniform? Secondly, why do the tables very often show this property? While the first question was discussed by several mathematicians (e.g., *R. S. Pinkham* and *J. H. B. Kempermann*) with fairly good solutions, the answers to the second one are not satisfactory. They frequently lead to confused "philosophies" even to number mysticism. According to Benford, e.g., while "Man" counts arithmetically: 1, 2, 3, ..., "Nature" automatically takes the logarithm of numbers and counts $e^0, e^x, e^{2x}, \ldots$. Benford states that the data of nature are composed of geometrical series for which (just like for the powers of 2) Benford's law holds. He also mentions several examples in the fields of science and technology which show the influence of Fechner's law discovered in the 19th century. According to this law, the relation between stimulus and sensation is logarithmic. Unfortunately, Benford's analogies do not give a satisfactory answer to the question. Further details can be found in *R. A. Raimi*'s survey article, where the author gives a very detailed reference.

d) Remarks

(*i*) If we are asked for the probability p_k that k ($k=1, 2, \ldots, 9$) is the first digit of a random entry from a table of numerical data and we suppose the existence of a definite solution and its scale invariance (nothing has been said about the scale units employed) then we arrive at the log-uniform distribution: $p_k = \lg(k+1) - \lg k$.

(*ii*) Analyzing the second or third, etc. digits, we realize that the influence of the "Benford's effect" can hardly be seen, if at all, i.e., the second, third, etc. digits are approximately uniformly distributed.

e) References

Benford, F., "The law of anomalous numbers", *Proc. Amer. Phil Soc.*, **78**, 551—572, (1938).

Newcomb, S., "Note on the frequency of use of the different digits in natural numbers", *Amer. J. Math.*, **4**, 39—40, (1881).

Raimi, R. A., "The first digit problem", *The American Math. Monthly*, **83**, 521—538, (1976).

8. THE PARADOX OF ZERO PROBABILITY; (CAN MANY NOTHINGS MAKE SOMETHING?)

a) The history of the paradox

The probability of an impossible event is zero, but the contrary is not true: the probability of an event may be zero without its being impossible. For example, the probability that we hit the very centre of the target is 0, though it is not impossible. There is also a zero probability of hitting any of one thousand fixed points, though this seems more likely than hitting the center point. Therefore the question arises of whether it is possible to compare the "chance" of events with zero probability or not. The other problem is that the probability of hitting a particular point of the target is zero, but a marksman will certainly hit one of the points, so the union of events with probability zero may be an event with probability one, that is many nothings can really make something. Is this, in fact, possible? This paradox is similar to Zenon's two and a half thousand year old paradox about the impossibility of moving. Zenon said that a flying arrow is still in every instance (or in other words, the displacement of the arrow during time intervals of zero length must be also zero), so it is unthinkable that it moves at all. The question is the same: how is it possible that adding many "nothings" result in "something"? Thus the essence of our paradox is several thousand years old, but its satisfactory explanation evolved only in the past decades, due to the researches of *Abraham Robinson* (1918—1974).

b) The paradox

We choose a point at random in the interval (0, 1). Then the probability that we have chosen exactly the point 1/2 is zero, just as the probability that we have chosen any of the points 1/100, 2/100, 3/100, ... though the latter seems to be more likely. Is it really impossible to make a difference between the probabilities of the two events?

c) The explanation of the paradox

In the history of arithmetic more and more complex types of numbers have been introduced: natural numbers and fractions were followed by the zero, the negative, the real (=rational+irrational) and the complex numbers. In the nineteensixties the set of numbers were further extended by introducing infinitely small numbers, the infinitesimals. The word "infinitesimal" itself had been used since the time of Newton and Leibniz in differential and integral calculus, but only symbolically without a well defined meaning or foundations. For precisely this lack of foundations, infinitesimals were expelled from rigorous mathematics in the past century, but they did not disappear completely (as physicists used them continuously). Mathematicians changed over to the use of "epsilon-delta" analysis, and this still describes the spirit of university educations. Robinson's theory, however, builds a firm logical foundation "under" the use of infinitesimals and the students of the next century will probably be taught in the revived spirit of Newton's and Leibniz's original heuristics. (At the universities of Wisconsin and M. I. T., students can, if they like, choose Robinson's theory instead of the epsion-delta theory of Weierstrass.) Infinitesimals can usually be used in calculations similarly to other numbers. While division by zero is not allowed, the division by an infinitesimal is well defined: the reciprocal value of an infinitesimal is an infinitely large number, and conversely, the reciprocal value of an infinitely large number is always an infinitesimal. Before Robinson's theory we thought that rational and irrational (i.e., real) numbers entirely fill the number line. Examining one point of the number line under Robinson's "mathematical microscope", we see not only one point but a multitude of infinitesimals which are infinitely near to this point.

This image is called a "monad" out of respect for Leibniz. Many paradoxes can be solved by the infinitesimals, Zenon's paradox as well as the paradox of zero probability. The point is that we have to make a difference between zero and infinitesimally small numbers. It is possible, for example, to assign a probability to every subset of an interval and this probability is zero only for the empty subset corresponding to the impossible event, and any other event will have positive, possibly infinitesimal, probability. Furthermore, considering a set A, whose probability is $P(A)$ in traditional sense, will have a probability which differs from $P(A)$ by at most an infinitesimal. (This new probability is not sigma-additive, only additive.) Now we may really say that the probability of choosing a single point, e.g., the centre of an interval is smaller than the probability of choosing one of two points: the difference is an infinitesimal.

d) Remark

Newton endeavoured to put the laws of nature into mathematical form, thus he arrived at the border between finite and infinite quantities, whereas Robinson apprehended infinity itself (having followed the example of *G. Cantor* and others), and made it familiar to everyday mathematics.

e) References

Luxemburg, W. A. J. (ed.), *Applications of Model Theory to Algebra, Analysis and Probability*, Holt, Rinehart and Winston, New York, 1969.

Robinson, A., "Non-standard analysis", *Proc. Nederl. Akad. Wetensch.*, **64**, 432—440, (1961).

Robinson, A., *Non-standard Analysis*, North-Holland, Amsterdam, 1966.

9. THE PARADOX OF INFINITELY DIVISIBLE DISTRIBUTIONS

a) *The history of the paradox*

The notion of infinitely divisible distributions was introduced by *B. de Finetti* in 1929. A distribution F is called infinitely divisible if for any positive integer n there exist n independent, identically distributed random variables such that the distribution function of their sum is just F. Let F_1 and F_2 be the distribution function of two independent random variables, and denote the distribution function of their sum by $F_1 * F_2$. The operation $*$ is called convolution. Obviously

$$(F_1 * F_2) * F_3 = F_1 * (F_2 * F_3), \quad F_1 * F_2 = F_2 * F_1$$

(which means algebraically that the distribution functions with the convolution as operation form a commutative semigroup). The distribution function F is $*$-infinitely divisible (by the above definition) if for every natural number n there exists a distribution function F_n such that

$$\underbrace{F_n * F_n * \ldots * F_n}_{n \text{ times}} = F.$$

Among others, the normal, Poisson, and exponential distributions are infinitely divisible. The most important role of these distributions is that they appear as limit distributions of the sum of independent random variables. In 1936 *Cramér* proved that if the convolution of two distributions is normal then both distributions must be normal and thus infinitely divisible. Two years later the same result was obtained by *Raikov* for Poisson distributions. These results are surprising since they claim that normal distributions can only be a decomposed into normal ones and the same stands for Poisson distributions. What is even more surprising is that infinitely divisible distributions can be decomposed into components which are not infinitely divisible.

b) *The paradox*

There exist distribution functions which are not infinitely divisible, but their convolution is infinitely divisible.

c) The explanation of the paradox

We will show that the exponential distribution can be decomposed into the convolution of distributions which are not infinitely divisible. Consider the exponential distribution with parameter $\lambda=1$, its density function is e^{-x} if $x>0$ and 0 if $x<0$. This distribution is really infinitely divisible since the nth convolution power of the gamma-distribution of order $1/n$ (whose density function is

$$\frac{x^{1/n-1}e^{-x}}{\Gamma\left(\frac{1}{n}\right)} \quad \text{if} \quad x > 0$$

and 0 if $x<0$) is just our exponential distribution. However, it can be decomposed not only into gamma-distributions (which themselves are infinitely divisible, too), but also into the convolution of two distributions one of which takes only the values $k=0, 1, 2, \ldots$ with probability $2^{-(k+1)}$, and the other is concentrated to the interval $(0, 1)$, i.e., it takes values from $(0, 1)$ .The latter distribution is not infinitely divisible. According to Remark (*i*), no distribution of bounded random variables can be infinitely divisible. So we know already that an infinitely divisible distribution may have not infinitely divisible convolution components. Now both components of the exponential distribution obtained above can be decomposed further so that each component concentrate to two points, more precisely, to 0 and an integer power of 2. These distributions (as every distribution concentrated to two points) are not only non-infinitely divisible, but just the contrary, they are irreducible (i.e., there is no way to decompose them unless one of the components is degenerated, i.e., concentrated to a single number).

d) Remarks

(*i*) We will show that the distribution function of a bounded random variable cannot be infinitely divisible unless the random variable is degenerated (i.e., when it takes only a single value with probability 1; in this case its variance is 0). If the random variable X is bounded then there exists a number K such that $|X|<K$. If the distribution function of X

is infinitely divisible, there are independent random variables $X_1, X_2, \ldots$ $\ldots, X_n$ with the same distribution such that the distribution functions of $X_1+X_2+\ldots+X_n$ and X are identical. As the supremum and variance of the sum of independent random variables is the sum of the suprema and variances of the components, we have

$$|X_i| < K/n \quad \text{and} \quad D(X_i) = D(X)/\sqrt{n}.$$

Consequently, if $D(X)\neq 0$ and n is large enough then the variance of X_i would be greater than the supremum of $|X_i|$, which is impossible. Therefore, if X is bounded and infinitely divisible then $D(X)=0$, i.e., X is degenerated.

(*ii*) Besides the normal, Poisson, and gamma-distributions, the log-normal distribution is also infinitely divisible. (It is the distribution of a positive random variable the logarithm of which is normally distributed—see Thorin's paper). Student's t-distribution and the Cauchy-distribution (the distribution of the quotient of two independent standard normally distributed random variables) are also infinitely divisible (see Lukacs's book and the papers by Grosswald, Epstein and Bondesson).

(*iii*) The exponential distribution serves as an example for infinitely divisible distributions decomposable into a (countably) infinite convolution of irreducible distributions. It is even more surprising that there exist infinitely divisible distributions which can be decomposed into the convolution of only two irreducible distributions (see Lévy's paper).

(*iv*) Infinitely divisible distributions were characterized by *Kolmogorov, Lévy* and *Hinčin* in the 1930es. It is easy to show that the distribution function of $X_1+X_2+\ldots+X_N$ is always infinitely divisible if $X_1, X_2, \ldots$ are arbitrary nonnegative integer valued, independent, identically distributed random variables and N is a Poisson distributed random variable independent of all X's. At the same time it follows from Lévy's and Hinčin's theorem that every infinitely divisible distribution concentrated to the nonnegative integers must be of this kind. Despite the fact that characterization theorems of infinitely divisible distributions are already 50 years old, the problem of characterization of infinitely divisible distribution having only infinitely divisible convolution components (the normal and the Poisson distributions belong to this class but, as we have seen, the exponential distribution does not) is even now an unsolved problem.

(*v*) In probability theory the notion of infinite divisibility comes up not only in connection with convolution. Other important operations can also be defined between distribution functions. For example, the distribution function of the maximum of two independent random variables with distribution functions $F_1(x)$ and $F_2(x)$ is $F_1(x) \cdot F_2(x)$. Therefore the product of distribution functions often appears in many probabilistic problems, e.g., in reliability theory when we want to obtain the probability distribution of life times in shunt connection. Obviously, for any natural number n and any distribution function $F(x)$, $\sqrt[n]{F(x)}$ is also a distribution function, thus every (one-dimensional) distribution function is infinitely divisible. In the case of more than one dimension, the characterization of infinitely divisible distributions is less trivial (see the paper by Balkama and Resnick). A third operation is the following.

Let $F_1 \circ F_2$ denote the multiplicative convolution, i.e., the distribution function of the product of independent random variables with distribution functions F_1 and F_2. While the Poisson distribution is infinitely divisible if the operation is the convolution $*$, it is not divisible if the operation is $\circ$. Moreover, if X and Y denote independent random variables and XY is of Poisson distribution, then either X or Y is concentrated to the two-element set $\{0, 1\}$ with probability 1. This means that the Poisson distribution is $\circ$-irreducible (see the paper by Székely and Zempléni). At the same time, the standard normal distribution is $\circ$-infinitely divisible, too. (The $\circ$-infinite divisibility of normal distributions with positive expectation is not yet proved or disproved; if the expectation is negative, it is obviously not $\circ$-infinitely divisible.)

e) References

Balkama, A. A., Resnick, S. I., "Max-infinite divisibility", *J. Appl. Prob.*, **14**, 309—319, (1977).

Bondesson, L., "A general result of infinite divisibility", *Annals of Probability*, **7**, 965—979, (1979).

Epstein, B., "Infinite divisibility of Student's t-distribution", *Sankhya, Ser. B.*, **39**, 103—120, (1977).
Fisz, M., "Infinitely divisible distributions: recent results and applications", *Annals of Math. Statist.*, **33**, 68—84, (1962).
Göndőcs, F., G. Michaletzky, T. Móri, G. J. Székely, "A characterization of infinitely divisible Markov chains with finite state space", *Ann. Univ. Sci. Budapest Sect. Math.*, **27**, 137—141, (1985).
Grosswald, E., "The Student t-distribution of any degree of freedom is infinite divisible", *Zeitsch' Wahrsch' theorie verv. Geb.*, **36**, 103—109, (1976).
Lévy, P., "Sur les exponentielles de polinómes", *Ann. Sci. École Normale Supérieure.* **54**, 231—292, (1937).
Lukacs, E., *Characteristic Functions*, Griffin, London, 1960.
Steutel, F. W., "Infinite divisibility in theory and practice", *Scand. J. Statist.*, **6**, 57—64, (1979).
Székely, G. J. "Multiplicative infinite divisibility of the normal distribution", *Proc. 7th Brasov Conf. on Probab. Theory*, Acad. Publ., Bucureşti, 579—582, 1984.
Székely, G. J., Zempléni, A., "Advanced problem 6431", *The American Math. Monthly*, **90**, 402, (1983).
Székely, G. J., "Problem 180." *Statistica Neerlandica*, **39**, 324, (1985).
Thorin, O., "On the infinite divisibility of the lognormal distribution", *Scand. Acturial J.*, 121—148, (1977).
Zolotarev, V. M., "On a general theory of multiplication of distributions of independent random variables", *Dokl. Acad. Sci. USSR*, **132**, 388—389, (1962), (in Russian).

10. PARADOXES OF CHARACTERIZATION

a) The history of the paradox

The originator of the following problem is again *George Polya*. Consider two independent identically distributed random variables X and Y. Is it possible that $aX+bY$ has the same distribution as X and Y if a and b are positive numbers? Polya analyzed this question in his paper published in 1923. The next remarkable result appeared only after a long interval in 1936, when *E. C. Geary* began to describe the distributions F that had the following property: if the variables $X_1, X_2, \ldots, X_n$ are independent and follow the distribution F, then

$$\bar{X} = \frac{X_1+X_2+\ldots+X_n}{n} \quad \text{and} \quad S = \sum_{i=1}^{n} (X_i-\bar{X})^2$$

are also independent. *M. Kac* in 1939 and *S. N. Bernstein* in 1941 answered the question: if X and Y are independent and identically distributed variables, then under what conditions are the variables $X+Y$ and $X-Y$ independent? Since the forties, characterization has evolved into a very important part of probability theory, both theoretically and practically, after the work of such outstanding mathematicians as *Yu. V. Linnik, E. Lukacs, A. A. Zinger, C. R. Rao and A. M. Kagan.*

b) The paradoxes

Let $X_1, X_2, \ldots, X_n$ be independent, identically distributed random variables. Is it possible, that $Y=f(X_1, X_2, \ldots, X_n)$ and $Z=g(X_1, X_2, \ldots, X_n)$ are identically distributed, or independent if f and g are different—e.g., linear—functions? In certain cases, for example, if f (and consequently Y) is identically constant, then Y and Z are obviously independent no matter what the function g is, but "generally" we would expect that Y and Z are neither independent nor identically distributed. Surprisingly, however, exceptions turn up precisely in the most important cases, when the X_i's are normally distributed. If, for example, X_1 and X_2 represent the coordinates of the velocity vector of a point moving randomly in a plane, and X_1, X_2 are independent standard normal variables, then the quantities $Y=X_1^2+X_2^2$, (which is proportional to the kinetic energy) and $Z=X_1/X_2$, (which determines the direction of the motion) are independent. These kinds of properties often characterize normal distributions (or other important distributions).

c) The explanation of the paradoxes

Let both f and g be linear functions:

$$Y = \sum_{i=1}^{n} a_i X_i \quad \text{and} \quad Z = \sum_{i=1}^{n} b_i X_i .$$

If there exist numbers a_i and b_i such that $a_i b_i$ is not always zero, and $a_i = b_i$ does not hold for every i, but Y and Z are still identically distributed and all the moments of X_i's are finite, that is, $E(|X_i|^k)$ is finite for every k, ($k=0, 1, 2, \ldots$), then the X_i's are normally distributed. This theorem of *J. Marcinkiewicz* is a generalization of Polya's theorem. (For further generalization we refer to the book by Kagan, Linnik and Rao.)

The theorem of *G. Darmois* and *V. R. Skitovič* states that if

$$Y = \sum_{i=1}^{n} a_i X_i \quad \text{and} \quad Z = \sum_{i=1}^{n} b_i X_i$$

are independent and $a_i b_i$ does not equal zero for every i, then X_i's are normally distributed.

The following generalization of Geary's theorem (also involving nonlinear functions) is very important in mathematical statistics. It states that if $\overline{X}$ and S are independent, and $n \geqq 2$, then the X_i's are normally distributed.

d) Remarks

(*i*) While the independence of $\overline{X}$ and S is a very strong condition, their correlation coefficient $r(\overline{X}, S)=0$ e.g., for all symmetric random variables $X_1, X_2, \ldots, X_n$ when the correlation exists. Though $r(\overline{X}, S)$ is always less than 1, its supremum is 1. For unimodal distributions the sharp upper bound is $\sqrt{15/16}$. In proving this result one can apply the sharp inequality $m_3^2 \leqq (m_4 - m_2^2) m_2$ where $m_k = E(X-E(X))^k$, $k=2, 3, 4$.

(*ii*) There are several interesting and natural ways of characterizing a family of distributions. For example, exponential distributions can be characterized by the following property: the entropy

$$-\int f(x) \log_2 f(x)\, dx$$

[$f(x)$ denotes the density function] is maximal for the exponential distributions among all distributions on the interval $(0, \infty)$ which have a given expectation. Among distributions on the interval $(-\infty, \infty)$ with given expectation and variance normal distributions have maximal entropy. On a finite interval the entropy is maximized by uniform distributions (without any further assumption).

e) References

Galambos, J., Kotz, S., *Characterizations of Probability Distributions*, Springer, Berlin—Heidelberg—New York, 1978.

Kagan, A. M., Linnik, Yu, V., Rao, C. R., *Characterization Problems in Mathematical Statistics*, Wiley, New York, 1973.

Mathai, A. M., Pederzoli, G., *Characterizations of the Normal Probability Law*, Wiley E. L., New Delhi—Bangalore—Bombay, 1977.

Polya, G., "Herleitung des Gauss'schen Fehlergesetzes aus einer Funktionalgleichung", *Math. Zeit.*, **18**, 96—108, (1923).

11. PARADOXES OF FACTORIZATION

a) The history of the paradox

The basic theorems in classical probability theory (such as the laws of large numbers, theorems on limit distribution) concern the distribution of the sum of independent random variables on the basis of the properties of its terms. The "converse" of these "composition" theorems are the "decomposition" or "factorization" theorems where the distribution of the sum is known and we want to gain some information on the possible terms or "factors". Such a decomposition result is *Cramér*'s theorem which has already mentioned. It claims that all factors of a normal distribution are also normal. Both in composition and decomposition theorems the characteristic function of random variables plays an important "technical role". The characteristic function of a random variable X is defined as the expectation of the complex random variable e^{itX} ($i=\sqrt{-1}$ and t is a real number), i.e., $\varphi_X(t)=E(e^{itX})$. Every random variable has a characteristic function, which uniquely determines the distribution function of the variable. The characteristic function of the sum of independent random variables is the product of the characteristic functions of the terms. These properties make it clear why characteristic functions are so extremely important for the solution of composition and factorization problems. They were already used by *A. Cauchy* in 1853 and *A. M. Lyapunov* at the turn of the century. Since the 1920s, mainly due to the work of *G. Polya* and *P. Lévy,* charactersitic functions have been used very frequently in solving composition problems. Since the 1930s, due to the theorems of *Cramér, Hinčin* and *Raikov,* the theory of decomposition has also evolved. There is no lack of surprising results or paradoxes in this field either (see the ones below).

b) The paradoxes

(*i*) There exist random variables X, Y and Z such that the probability distribution of $X+Y$ is equal to that of $X+Z$ but the distribution of Y and Z are not the same. This fact, first pointed out by *Hinčin* in 1937, is rather surpising because if X is either a bounded random variable or its characteristic function is never equal to 0 (e.g., if it is infinitely divisible) then the distributions of Y and Z must also be equal. Owing to Hinčin's paradox, in general there is no sense to speak about the "rest" of a probability distribution after one of its factors has been cancelled because the remaining part is not unique. A great many difficulties are caused by this fact in the algebra of probability distributions. At the same time it is reasonable to ask what remains if a normal factor of a distribution is omitted, since the characteristic function of a normal distribution is never equal to 0 (the characteristic function of the standard normal distribution is $e^{-t^2/2}$). A certain degree of caution, however, is required even in this case. Namely, there exist independent, identically distributed random variables X and Y that have no normally distributed factors (with positive variance) but the random variable $X+Y$ has already got one. This result was first pointed out by *D. Dugué* and *R. A. Fisher* in 1948 (see their paper below).

(*ii*) Let $X_1, X_2, \ldots, X_n$ be independent random variables whose distributions (not necessarily the same) are not known. We know, however, the distributions of the linear combinations

$$Y_j = \sum_{k=1}^{n} c_{jk} X_k, \quad j = 1, 2, \ldots, n;$$

(c_{jk} is an arbitrary number). If there exist $X_1, X_2, \ldots, X_n$ satisfying this system of equations and the determinant of the matrix (c_{jk}) is not 0, then (since in this case $Y_1, Y_2, \ldots, Y_n$ determine $X_1, X_2, \ldots, X_n$ uniquely) we might think that the distributions of $X_1, X_2, \ldots, X_n$ are also uniquely determined. As *A. Rényi* showed in 1950, this is not the case.

c) The explanation of the paradoxes

(*i*) It can be shown that if the value of a function $\varphi(t)$ is nonnegative for any real t, $\varphi(0)=1$, $\varphi(-t)=\varphi(t)$ and for positive arguments t the function $\varphi(t)$ is convex and $\lim \varphi(t)=0$ as $t\to\infty$, then there exists a random variable whose characteristic function is $\varphi(t)$. Thus there is a random variable X with characteristic function $1-|t|$ if $|t|\leqq 1$ and 0 otherwise, and there also exist random variables Y and Z whose characteristic functions are the same on the interval $|t|\leqq 1$ but differ outside this interval. Therefore if X, Y and Z are independent then

$$\varphi_{X+Y}(t) \equiv \varphi_{X+Z}(t),$$

i.e., $X+Y$ and $X+Z$ have the same distribution, while the distributions of Y and Z are different. (We shall see another example in 13a).

Slightly more can be proved than stated in the paradox. It can be shown that when $\psi(t)$ is a periodical function with period length of 2 and $\psi(t)=1-|t|$ whenever $|t|\leqq 1$, then there exists a random variable Y whose characteristic function is just $\psi(t)$. From this it follows that $\varphi(t)\psi(t)\equiv\varphi(t)^2$, i.e., there are independent random variables X, Y and Z such that the distribution of $X+Y$ is equal to that of $X+Z$, and the distributions of X and Z are the same while those of Y and Z differ.

(*ii*) If the random variables Y_j are given and the determinant $\|c_{jk}\|\neq 0$ then the random variables X_j are uniquely determined. The variables Y_j can, however, be given in several different ways so that their distributions remain unchanged. Therefore it is not at all certain that the distributions of Y_j uniquely determine the distributions of X_j if only $\|c_{jk}\|\neq 0$ is supposed.

d) Remarks

Rényi proved that generally it is also necessary to suppose that $\|c_{jk}^2\|$ is not equal to 0. If we know that

$$\|c_{jk}\| \neq 0 \quad \text{and} \quad \|c_{jk}^2\| \neq 0$$

then under general conditions the uniqueness is also guaranteed (e. g., if the characteristic function of Y_j is an entire function of order $\leqq 2$

and if there is a solution at all). This fact has very important practical consequences. We have seen in II. 13/c that two values can be obtained by two measurements more accurately (the variances are less) if they are not measured one by one but if first their sum, then their difference is measured. The case is similar if we want to know n different unknown quantities $X_1, X_2, \ldots, X_n$. Greater accuracy can be achieved if certain linear combinations $Y_1, Y_2, \ldots, Y_n$ are measured. Generally it is reasonable to use such a matrix (c_{jk}) which consist of $+1$ and -1 only. In this case obviously $\|c_{jk}^2\|=0$, and therefore the distributions of $Y_1, Y_2, \ldots$ $\ldots, Y_n$ do not determine uniquely the distributions of $X_1, X_2, \ldots, X_n$. As an example, let the distribution of both $Y_1=X_1+X_2$ and $Y_2=X_1-X_2$ be standard normal. Then Cramér's theorem states that X_1 and X_2 are also normally distributed with expected value 0. However, their variance is not determined uniquely. They merely satisfy the relation $D^2(X_1)+D^2(X_2)=1$.

e) References

Dugué, D., Fisher, R. A., "Un résultat assez inattendu d'arithmétique des lois des probabilités. *C. r. Acad. Sci.*", **227**, 1205—1207, (1948).

Feller, W., *An Introduction to Probability Theory and its Applications,* Vol. II. Wiley, New York, 1966 (Chapter XV).

Lukacs, E., *Characteristic Functions,* Griffin, London, 1960.

Rényi, A., "On the algebra of distributions", *Publ. Math.,* **1**, 135—149, (1950).

12. THE PARADOX OF IRREDUCIBLE AND PRIME DISTRIBUTIONS

a) The history of the paradox

Irreducible numbers, i.e., integers (greater than one) that have only one and themselves for divisors, play a fundamental part in arithmetic. These numbers: 2, 3, 5, 7, 11, 13, 17, 19, ... are also *prime numbers,* i.e., if they are factors of the product of two natural numbers, then they are also factors of at least one of these numbers. Among natural numbers primes and irreducibles are the same and the numbers 2, 3, 5, 7, ... are always called prime numbers. The fundamental theorem of arithmetic states that there is exactly one prime factorization of each integer greater than one (the order of the prime factors is disregarded). Thus prime numbers in arithmetic are like building blocks or like atoms in the physical world. The most natural way of getting information about a complicated structure is to break it down into atoms, hence it is understandable that the notions of irreducibility and primality have been extended to general algebraic structures (since the last century). These notions can also be interpreted for probability distributions: the role of natural numbers is taken over by probability distributions and the role of multiplication by convolution (for the definition of convolution see "The paradox of infinitely divisible distributions").

A distribution F is *irreducible* if $F=G*H$ implies that one of G and H is degenerated (i.e., concentrated to a single point with probability one; these distributions play the role of units). A distribution F is called a *prime distribution* if it is the factor of $G*H$ only if it is also the factor of G or H. Hinčin proved in 1937 that every distribution is the convolution of an infinitely divisible distribution and a finite or countable convolution product of irreducible distributions, i.e., every characteristic function $\varphi(t)$ can be expressed in the following form:

$$\varphi(t) = \psi(t) \prod_i \varphi_i(t),$$

where $\psi(t)$ and $\varphi_i(t)$ are the characteristic functions of infinitely divisible and irreducible distributions, respectively. This is somewhat similar to the fundamental theorem of arithmetic but there is an important difference: the factorization of distributions is not unique. If, for example the distribution F assumes the values 0, 1, 2, 3, 4, 5 with the same, 1/6, probability, then F can be decomposed to the convolution of irreducible distributions in two different ways: in the first decomposition the first factor assumes the values 0 and 1, the second factor assumes the values 0, 2 and 4 with the same probability; in the second decomposition the first factor assumes the values 0 and 3, and the second factor assumes the values 0, 1 and 2 with equal probabilities. This ambiguity shows that the analogy between the "arithmetic" of numbers and probability distributions is not perfect. The following paradox will show a more considerable discrepancy.

b) The paradox

Probability distributions with the operation of convolution form an algebraic structure that contains many irreducible distributions (e.g., every distribution concentrated on two points is irreducible), but it does not contain any prime distribution. So if we really regard primes as the "atoms" of probability distributions, then there are no atoms at all. *I. Z. Ruzsa* and *G. J. Székely* first pointed out this fact in 1979.

c) The explanation of the paradox

The fact that primes and irreducibles are usually different is not surprising at all, it seems unexpected only because in the most familiar and important structure, among natural numbers the two notions are equivalent. In general, however, we can only say that a prime is always irreducible but the opposite is not necessarily true. The coincidence of the two terms means (roughly speaking) that the factorization into irreducible elements is unique. It was observed even by Hinčin in 1937 that the convolution decomposition of probability distributions into irreducibles is not unique, that is, not every prime is irreducible. The article of Ruzsa and Székely shows that there are no primes in this structure at all, and thus concerning the connection between the notions of irreducibility and primality, the multiplicative structure of natural numbers and the convolution structure of probability distributions are opposite extremes.

d) Remark

Say that two distributions F and G are *relatively prime* if F and G can be the factors of a distribution H only if $F * G$ is also a factor of H. In the article already referred to, we proved that bounded (and not degenerated) distributions cannot be relatively prime, and we conjecture that there are no relatively prime distributions at all; but this problem is still unsolved.

e) References

Kendall, D. G., Harding, E. F. (eds.), *Stochastic Analysis,* Wiley, New York, 1973.

Linnik, Yu, V., Ostrovskii, I. V., *Decomposition of Random Variables and Vectors,* Translations of Mathematical Monographs, **48**, American Mat. Soc., Providence, 1977.

Ruzsa, I. Z., Székely, G. J., "No distribution is prime", *Zeitsch' Wahrsch' theorie verw. Geb.*, **70**, 263—270, (1985).

Ruzsa, I. Z., Székely, G. J. Decomposition in Semigroups—Algebraic Probability Theory, Wiley, New York (to be published).

Ruzsa, I. Z., Székely, G. J., "Theory of Decomposition in Semigroups", Advances in Math., **56**, 9—27, (1985).

Ruzsa, I. Z , Székely, G. J., "How to eliminate probabilities from probability theory?", (to be published).

Ruzsa, I. Z., Székely, G. J., "A note on our paper 'Theory of Decomposition in Semigroups'" *Advances in Math.*, (to be published).

Székely, G. J., Zempléni, A., "Multiplicative arithmetic of distribution functions", (to be published).

13. QUICKIES

a) The paradox of halving distributions

Let X and Y be independent, identically distributed random variables. The distribution of $X+Y$ usually determine uniquely the common distribution of X and Y, but, paradoxically, not always. This fact is surprising, because the distributions in practice can usually be divided uniquely, i.e., their nth proportion is determined uniquely (if it exists), as in the case of bounded or infinitely divisible distributions. Now let us see a paradox example.

If a random variable is restricted to the values $2k+1$ ($k=0, \pm 1, \pm 2, \ldots$), and the corresponding probabilities are

$$\frac{4}{\pi^2(2k+1)^2},$$

then its characteristic function $\varphi(t)$ is periodic with period 2π, and in the interval $-\pi \leqq t \leqq \pi$

$$\varphi(t) = 1 - \frac{2|t|}{\pi}.$$

We now define another random variable which assumes zero with probability 1/2 and assumes $4k+2$ with probability

$$\frac{2}{\pi^2(2k+1)^2}.$$

The characteristic function $\psi(t)$ of this random variable is also periodic. The length of its period is π, and in the interval $-\pi/2 \leqq t \leqq \pi/2$

$$\psi(t) = 1 - \frac{2|t|}{\pi}.$$

Obviously $\psi(t)=|\varphi(t)|$, so $\psi(t)^2=\varphi(t)^2$. Thus if the characteristic function of $X+Y$ is $\psi(t)^2=\varphi(t)^2$, then the common characteristic function of X and Y may be either $\varphi(t)$ or $\psi(t)$, so it is not uniquely determined.

We note that $(\psi(t)+\varphi(t))/2$ is also a characteristic function, thus

$$\frac{\varphi(t)+\psi(t)}{2}\varphi(t) = \frac{\varphi(t)+\psi(t)}{2}\psi(t),$$

which gives another example of the first factorization paradox, since this equation cannot be reduced by $(\varphi(t)+\psi(t))/2$.

One can also construct characteristic functions so that they do not always assume real values, but their square is always real. Consequently there exist probability distributions which are symmetric to the origin, but "half of them" are not symmetric in the sense that there exist independent, identically distributed random variables X and Y such that $X+Y$ is symmetrically distributed, but X and Y are not. (In the case of bounded variables this may not happen.) It also seems surprising that if the random variable X has a symmetric density function $f(x)$ $(f(-x)=f(x))$ such that $0<a<f(x)<b<\infty$ whenever $|t|\leqq c<\infty$ and $f(t)=0$ for $|t|>c$, than X has no half, i.e., there is no characteristic function φ such that $\varphi_X=\varphi^2$. (Ref.: Problem 10, *Mat Lapok*, **30**, 1982, p. 272. (in Hungarian). Proposed by T. F. Móri and G. J. Székely.)

b) Pathological probability distributions

(*i*) Let the distribution function F of a random variable have the following properties: $F(0)=0, F(1)=1$ and for

$$x = \sum_{r=0}^{\infty} 2^{-a_r}, \quad (a_0 < a_1 < a_2 < \dots \text{ positive integers})$$

$$F(x) = \sum_{r=0}^{\infty} a^r (1+a)^{-a_r},$$

where a is an arbitrary positive number. *L. Takács* showed (*The American Math. Monthly*, **85**, 35—37, 1978) that $F(x)$ is a strictly monotone increasing and continuous function on the interval (0, 1). For $a=1$, $F(x)=x$ on the interval (0, 1), that is, the random variable is uniformly distributed, its density function is zero except the interval (0, 1), where it is one. Surprisingly, if $a\neq 1$, then F (and the corresponding random variable) never has a density function, that is, there does not exist a function f, for which

$$F(x) = \int_{\infty}^{x} f(u)\, du.$$

Though the most frequent continuous probability distributions always have density functions, we must not forget about the pathological random variables we have just mentioned. It is interesting, for example, that if a uniformly distributed random variable is decomposed into the sum of two independent random variables whose distribution functions are continuous, then at least one of them is pathological, i.e., it does not have a density function. (The history of pathological and very pathological, i.e., "singular" functions began in 1904, when *H. Lebesgue* published his book on integration of functions. One of the latest results on singular functions is due to *T. Zamrifescu*, "Most monotone functions are singular", *The American Math. Monthly*, **88**, 47—49, (1981). See also F. S. Cater, "Most monotone functions are not singular", *The American Math. Monthly*, **89**, 466—469, (1982).)

(*ii*) Let be the joint density function of X and Y be

$$h(x, y) = \frac{|x|}{2\sqrt{2\pi}} e^{-(|x|+x^2y^2/2)}.$$

Then the density function of X is

$$f(x) = \int_{-\infty}^{+\infty} h(x, y)\, dy = \begin{cases} 0, & \text{if } x = 0 \\ \frac{1}{2} e^{-|x|}, & \text{if } x \neq 0. \end{cases}$$

Clearly h is continuous, but f is not at zero. *L. E. Clarke* constructed an example (*The American Math. Monthly,* **82**, 845—846, (1975)) where h is continuous everywhere, but f is nowhere continuous! It can easily be shown that f is always lower semi-continuous, i.e.,

$$\lim_{x' \to x} f(x') \geqq f(x).$$

Moreover, if the integral of a non-negative, semi-continuous function extended over the entire x axis is unity, then there exists a continuous density function $h(x, y)$, such that

$$f(x) = \int_{-\infty}^{\infty} h(x, y)\, dy.$$

(Ref.: Pelling, M. J., Verbeek, A., "On marginal density functions of continuous densities II", *The American Math. Monthly*, **84**, 364—365, (1977).)

c) The newsagent paradox

A newsagent orders N dailies every day. He makes a profit of b dollars on every daily sold and has a loss of c dollars on every daily left over. Which N should he choose to maximize his expected profit? The numbers of customers naturally depends on chance. Suppose it follows the Poisson distribution with some parameter λ, that is, the probability that the number of customers is exactly n is $\lambda^n e^{-\lambda}/n!$. If we put $b=1$, $c=2$ and $\lambda=10$, i.e., the average number of customers is 10, then one can show that the average number of dailies that should be ordered is 9. It is evident, however, that if the newsagent orders only 9 dailies every day, the average number of customers will decrease from 10 to 9; but in this case the optimal number of dailies would only be 8, etc. The explanation of this paradoxical situation lies in the fact that we must take into account the loss, caused by losing a "potential customer", who leaves disappointedly.

Let d dollars be the loss of the newsagent if he cannot serve a customer with a daily. (The value of d cannot be determined as simply as the values of b and c, hence—unfortunately—it is often neglected. For example, if $d=1\$$, for $\lambda=10$ the optimal value of N is 10.)

In general denote by X the (random) number of customers (now do not suppose that X is Poissonian). One can show that the optimal number of N is the solution of the equation $P(X>N)\approx\frac{c}{b+c+d}$, i.e. if the newsagent stocks N copies of a daily paper where N is the solution of this equation, then his expected profit will be maximized.

(Ref.: Morse, P. M., Kimball, G. E., *Methods of Operations Research,* Wiley, New York, 1951.
DeGroot, M. H., *Optimal Statistical Decisions,* McGraw Hill, New York, 1970.)

d) Kesten's paradox

According to Kolmogorov's strong law of large numbers, if $X_1, X_2, X_3, \ldots$ are independent, identically distributed random variables, then the sequence

$$\overline{X}_n = \frac{X_1+X_2+\ldots+X_n}{n}$$

converges to a constant M with probability one, if and only if the expectations of X_i's exist; then these expectations are just equal to M. Thus if M exist, then the sequence $\overline{X}_n$ is very "regular" with probability one: its only accumulation point is M. To what extent can the sequence $\overline{X}_n$ be "irregular" if M does not exist? *Harry Kesten* proved in 1970 that the set of the limit points of the sequence $\overline{X}_n$ can be an arbitrary closed set (independent of chance) which contains $-\infty$ and ∞ with probability one. Accordingly, the set of limit points may be the entire number line, though it has not yet been established what kind of characteristics the distribution function of X_i's must have to possess this property.

(Ref.: Kesten, H., "The limit points of a normalized random walk", *Annals of Math Statist.,* **41**, 1173—1205, (1970).)

e) The paradox of the stochastic geyser

Alfred Rényi proposed the following question in 1962. Consider a geyser which is gushing at intervals $X_1, X_2, \ldots$; suppose these are independent, identically distributed random variables. We measure the times of gushes, $S_n = X_1 + X_2 + \ldots + X_n$. How large can the measuring errors be if we want to determine the unknown distribution of X_i's with probability one, on the basis of our measurement data. This question is in close connection with the following problem. Let $S_n = X_1 + X_2 + \ldots + X_n$ and $T_n = Y_1 + Y_2 + \ldots + Y_n$ be the partial sums of independent, identically distributed random variables (X's need not be independent of Y's). How precisely can S_n approach T_n if the distributions of X's and Y's are different? According to the strong law of large numbers, if $(S_n - T_n)/n$ tends to zero, then the expected values of the X's and Y's (provided that they exist) cannot be different (with probability one). The well-known law of the iterated logarithm shows that if the standard deviations $D(X)$ of X's exist, then

$$\limsup_{n \to \infty} \frac{S_n - E(S_n)}{\sqrt{2n \ln \ln n}} = D(X).$$

Consequently, if even $(S_n - T_n)/\sqrt{n}$ converges to zero, then the standard deviations of X's and Y's must be equal. The researches of Skorohod and Strassen—based on the theory of Brownian motion—led to the conjecture that if, e.g., the X's are bounded and the Y's are normally distributed, then $|S_n - T_n|$ is at least $\sqrt[4]{n}$, (if n is large enough). Hence $(S_n - T_n)/\sqrt[4]{n}$ cannot converge to zero (with probability one). Relying upon these findings it was thought that the times of gushing of the stochastic geyser are enough to be measured with an error smaller than $\sqrt[4]{n}$. It was a great surprise that, after *P. Révész* and *M. Csörgő* had taken the initiative *J. Komlós, P. Major* and *G. Tusnády* showed in 1974 that S_n can be approximated by T_n so well that even $(S_n - T_n)/\ln n$ remains bounded. Thus if we record the times of gushes, the measurement errors must be kept within the limit of $\ln n$. P. Bártfai showed that if the measurement error divided by $\ln n$ converges to zero, then the distribution of intervals between subsequent gushes can be determined with probability one.

(Ref.: M. Csörgő, P. Révész, *Strong Approximations in Probability and Statistics*, Akadémiai Kiadó, Budapest, 1981.)

f) The paradox of probability in quantum physics

The methods of probability theory were widely used in physics even in the last century. Classical statistical physics started from the idea that the equilibrium of a system (consisting of large number of particles) is the most probable state of the system. The methods of statistical physics were thought to describe only approximately the macroscopic behaviour of a system. Through the probabilistic interpretation of quantum physics, however, chance and probability became a fundamental part of physics as a whole. Probability has become a basic notion such as energy, for example, not merely some kind of approximation which could be avoided in principle. Even Einstein was not pleased with this sweeping change in the foundations of physics, though he was not a bit conservative. He wrote in a letter to *Max Born* (who was awarded the Nobel Prize for his probabilistical interpretation of the quantum mechanical wave function) that he did believe in the existence of perfect laws of Nature: "God does not dice." In his answer, Born explained that instead of solving a great number of differential equations, in some cases one can obtain reasonable results by tossing dice. Since that time Born's conception has become dominant. Chance and probability are already accepted notions of physics. These changes also affected philosophy: mechanistic determinism lost its dominant importance. The present state of the world does not determine uniquely its future state. With our present knowledge we can determine only the probability of future events. This, however, does not mean agnosticism, since the laws of chance are recognizable (probability theory deals with exactly this). Paradoxically, the physical concept of probability is not simply the application of mathematical probability in physics. The motives and spirit of the two concepts are different. According to *R. P. Feynman,* who won the Nobel Prize for physics in 1965, the laws of quantum physics can be understood on the basis of probability theory that evolved from the theory of games of chance if we apply the laws of probability theory in the case of large numbers of particles, but these laws do not explain the behaviour of a *single* electron or proton. The wave theory of *de Broglie* and *Schrödinger* and the uncertainty principle of *Heisenberg* led to the elaboration of a new quantum-probability theory between 1926 and 1929, especially due to *Born.* Kolmogorov's mathematical theory of probability was also built up about that time. The clarification of the relation between the two kinds of probability theories began much later, about twenty years ago, especially with *G. Mackey*'s work based on some earlier research of von Neumann. At last a general and unified probability theory has developed, which involves both classical and quantum probability theory (cf. Gudder's book). This solved a contradiction and made it possible to outline a probability theory based on general event structures.

(Ref.: Born, M., *Natural Philosophy of Cause and Chance,* Dover Pub., New York, 1964;
Gudder, S. P., *Stochastic Methods in Quantum Mechanics,* North-Holland, New York, 1979.)

g) The paradox of cryptography

Throughout the several thousand years in the history of cryptography, cryptanalysts have invented more and more cunning ciphers, and their adversaries have correspondingly outwitted them by discovering more and more efficient techniques for craching ciphers. *Edgar Allan Poe,* who fancied himself a skilled cryptanalyst, was convinced that "... human ingenuity cannot concoct a cipher which human ingenuity cannot resolve". The first turning point in the history of cryptography was reached in the twenties, when "*one-time pads*" were discovered. These one-time ciphers were first used by the Germans and have been in general use for half a century. Different types of one-time pads are considered very efficient and are in constant use today in many countries, for special messages. The famous "hot-line" between Washington and Moscow also makes use of a one-time pad. These ciphers are really unbreakable in principle, since a different shift cipher is used to encode each symbol in the plaintext, each time choosing the shift at random. If the letter "e" was always encoded as "t", it would be a simple substitution cipher, easily broken by a statistical analysis (as "e" is the most frequent letter in many languages). If, however, "e" is encoded sometimes as "a", sometimes as "c" and sometimes as "w" and the substituting letter is chosen at random, then this one-time cipher is uncrackable even in principle. The ciphertext does not disclose anything about itself. The disadvantage of this procedure is that one-time pads are used only once, for a single message. A brilliant discovery of *Whitfield Diffie* and *Martin E. Hellman,* both electrical engineers at Stanford University, revolutionized the entire field of secret communication. Inspired by the mathematical theory of complexity they proposed a new kind of cipher in 1975 which is not unbreakable in principle but absolutely unbreakable in practice. More precisely, these new ciphers can be broken, but only by computer programs that run for millions of years. Surprisingly the encryption and decription procedure of Diffie and Hellmann are not symmetric, meaning that if only the method of encryption is known, it is computationally infeasible to discover the method of decoding, and this provides absolute secrecy. (This method of ciphering is made possible by what Diffie and Hellman call a trapdoor one-way function.) The secret can be locked and unlocked with different keys (and opening it requires a much finer key). The basic idea of one-way cipherment is very simple: two numbers can easily be multiplied by each other, e.g., the product of 101 and 211 can be calculated quickly, it is 21311; but if we want to find two integers greater than one, whose product is 21311, then it will take much more time to find that 101 and 211 is the only possible solution. Naturally there are computer algorithms for factoring numbers, but in the case of a 40—50 digit number, the running time required would be millions of years. On the basis of prime number theory, a simple trapdoor function was found: the enciphering key depends only on the product of two prime numbers, whereas to decipher the ciphertext the two prime numbers have to be known, too. Let us go into more details about this trapdoor function!

Let p and q be two, large, random prime numbers. The product n of these two numbers and another random number E are the user's enciphering key (E, n), which does not have to be kept secret; it can be put in a public file, such as a telephone directory. To apply the key, a sender first converts his message into a string of numbers, which he then breaks into blocks $B_1, B_2, \ldots$. Each plaintext number B_i must be between 0 and $n-1$. The sender computes for each plaintext number B_i the ciphertext number $C_i = B_i^E$ modulo n, (that is, if the Eth power of B_i is divided by n, the residual is C_i). This public-key cryptosystem is based on the fact that although finding large prime numbers (p and q) is computationally easy, factoring the product of two such numbers is at present computationally infeasible, so knowing only (E, n) and C_i, it is hopelessly difficult to find B_i. To decipher a ciphertext $C_1, C_2, \ldots$, the user employs n and a secret deciphering key D derived from the prime factors p and q of n. D is the multiplicative inverse of E modulo $(p-1)(q-1)$, that is, ED modulo $(p-1)(q-1)$ is equal to 1. (The product $(p-1)(q-1)$ is the number of integers between 1 and n that have no common factor with n.) After all these the receiver can easily obtain B_i:

$$C_i^D = (B_i^E)^D = B_i$$

mod n. This method was designed by *Rivest, Shamir* and *Adleman* and is called the RSA system.

(Ref.: Hellman, M. E., "The mathematics of public-key cryptography", *Sci. Amer.*, **241**, 130—139, (1979).
Simmons, G. J., "*Cryptology, the mathematics of secure communication*", *The Math. Intelligencer*, **1**, 233—246, (1979).
Shamir, A., "A polinomial time algorithm for breaking Merkle—Hellman cryptosystems", *Research Announcement*, 1982.)

h) The paradox of poetry and information theory

The last paradox in this book is a quotation from my late professor *Alfréd Rényi*.

"Since I started to deal with information theory I have often meditated upon the conciseness of poems; how can a single line of verse contain far more 'information' than a highly concise telegram of the same length. The surprising richness of meaning of literary works seems to be in contradiction with the laws of information theory. The key to this paradox is, I think, the notion of 'resonance'. The writer does not merely give us information, but also plays on the strings of the language with such virtuosity, that our mind, and even the subconscious self resonate. A poet can recall chains of ideas, emotions and memories with a well-turned word. In this sense, writing is magic."

Chapter 5

Paradoxology

> "On foundation we believe in the reality of mathematics, but of course when philosophers attack us with their paradoxes we rush to hide behind formalism and say, 'Mathematics is just a combination of meaningless Symbols,'..."
>
> (J. A. Dieudonné, 1970)

Like most branches of science, mathematics is also the history of paradoxes. The greatest discoveries generally solve the greatest paradoxes (think of Darwin or Einstein) while they serve as sources for new ones as well. Socrates' teaching method of perceiving new ideas through paradoxes is the most fundamental because the process of scientific cognition itself rests on paradoxes.

For the development of deductive mathematics it was of fundamental importance that (in spite of the Pythagorean "all is number", i.e., integer number) there are distances (e.g., the diagonal and the side of a square) whose ratio is not the ratio of integer numbers, which means that this ratio is not a number in Pythagorean sense. (In modern terminology, it is not a rational number.) This paradox of "incommensurability" led to the dissolution of the Pythagorean school and the overshadowing of number mysticism, to Euclidean geometry (where the role of numbers was replaced by geometric figures), and to Plato's "mathematical idealism" (in practice "incommensurability" cannot be tested directly, thus, according to Plato, experience cannot lead to real knowledge). The greatest paradox in the mathematics of the Middle Ages was that "nothing", i.e., nought should be considered something and denoted by a figure. In this way, due to the Indian-Arabic method of number writing, calculating became much easier. At the break of modern times several paradoxes were caused by negative and later complex numbers. E.g., one of them states that $(-1):1=1:(-1)$ is impossible because the ratio of a smaller number to a greater one cannot be equal to that of a greater number to a smaller one. Modern times have brought several new paradoxes in all branches of mathematics from the solvability of algebraic equation on to Bolyai's geometry. It is interesting that already in the first half of the last century B. Bolzano from Prague devoted a whole book to the paradoxes of infinity ("Paradoxien des Unendlichen") though the most interesting paradoxes of infinity appeared only after G. Cantor's set theory published in 1872. Most leading mathematicians of the century, such as Gauss, Cauchy, Kronecker, Poincaré and others, rejected the notion of actual infinity and assigned only symbolical meaning to it. The foundation stone of modern mathematics is, however, Cantor's theory using the notion of actual infinity, though we have to emphasize that the "horror infiniti" has not yet vanished. In fact new paradoxes have increased the number of finitists. Similarly, the fear of randomness is still in the air. The mathematical paradoxes of infinity and randomness are extremely important because these two concepts fundamentally influence our outlook and philosophical attitude. Probability theory has evolved as a symbolic counterpart of the random universe thus it is to be hoped that the paradoxes in this book will help the reader to find the best way through our random world.

"Rien ne m'est sûr que la chose incertaine;
Obscur, fors ce qui est tout evident;
Doute ne fais, fors en chose certaine;
Science tiens a soudain accident;"

(F. Villon, *Ballade Du Concours De Blois*, 11—14)

Notations

$n!$ $= 1 \cdot 2 \cdot 3 \cdots n$

$\binom{n}{k}$ $= \dfrac{n(n-1)(n-2)\ldots(n-k+1)}{1 \cdot 2 \cdots k}$

$P(A)$ probability of the event A

$P(\bar{A})$ probability of the complement of the event A $(\bar{A})$

$P(AB)$ probability of the joint occurrence of events A and B

$P(A|B)$ probability of the event A given that the event B has occurred

$\bar{X}$ $= \dfrac{X_1 + X_2 + \ldots + X_n}{n}$

$\hat{\vartheta}$ estimator of the parameter ϑ

$E(X)$ or EX expectation (or expected value) of the random variable X

$D(X)$ standard deviation of the random variable X

$\Gamma(z)$ $= \int_0^\infty e^{-t} t^{z-1} dt$ if the real part of the complex number z is positive;

$\Gamma(z+1) = z\Gamma(z)$, thus $\Gamma(n+1) = n!$; $\Gamma(1/2) = \sqrt{\pi}$

$\Phi(x)$ $= \dfrac{1}{\sqrt{2\pi}} \int_{-\infty}^{x} e^{-u^2/2} du$

Table 1. The standard normal distribution function

$$\Phi(x) = \frac{1}{\sqrt{2\pi}} \int_{-\infty}^{x} e^{-u^2/2}\, du \qquad [\Phi(-x) = 1 - \Phi(x)]$$

x	$\Phi(x)$	x	$\Phi(x)$	x	$\Phi(x)$	x	$\Phi(x)$	x	$\Phi(x)$
0,00	0,5000	0,51	0,6950	1,02	0,8461	1,53	0,9370	2,08	0,9812
0,01	0,5040	0,52	0,6985	1,03	0,8485	1,54	0,9382	2,10	0,9821
0,02	0,5080	0,53	0.7019	1,04	0,8508	1,55	0,9394	2,12	0,9830
0,03	0,5120	0,54	0,7054	1,05	0,8531	1,56	0,9406	2,14	0,9838
0,04	0,5160	0,55	0,7088	1,06	0,8554	1,57	0,9418	2,16	0,9846
0,05	0,5199	0,56	0,7123	1,07	0,8577	1,58	0,9429	2,18	0,9854
0,06	0,5239	0,57	0,7157	1,08	0,8599	1,59	0,9441	2,20	0,9861
0,07	0,5279	0,58	0,7190	1,09	0,8621	1,60	0,9452	2,22	0,9868
0,08	0,5319	0,59	0,7224	1,10	0,8643	1,61	0,9463	2,24	0,9875
0,09	0,5359	0,60	0,7257	1,11	0,8665	1,62	0,9474	2,26	0,9881
0,10	0,5398	0,61	0,7291	1,12	0,8686	1,63	0,9484	2,28	0,9887
0,11	0,5438	0,62	0,7324	1,13	0,8707	1,64	0,9495	2,30	0,9893
0,12	0,5478	0,63	0,7352	1,14	0,8729	1,65	0,9505	2,32	0,9898
0,13	0,5517	0,64	0,7389	1,15	0,8749	1,66	0,9515	2,34	0,9904
0,14	0,5557	0,65	0,7422	1,16	0,8770	1,67	0,9525	2,36	0,9909
0,15	0,5596	0,66	0,7454	1,17	0,8790	1,68	0,9535	2,38	0,9913
0,16	0,5636	0,67	0,7486	1,18	0,8810	1,69	0,9545	2,40	0,9918
0,17	0,5675	0,68	0,7517	1,19	0,8830	1,70	0,9554	2,42	0,9922
0,18	0,5714	0,69	0,7549	1,20	0,8849	1,71	0,9564	2,44	0,9927
0,19	0,5753	0,70	0,7580	1,21	0,8869	1,72	0,9572	2,46	0,9931
0,20	0,5793	0,71	0,7611	1,22	0,8888	1,73	0,9582	2,48	0,9934
0,21	0,5832	0,72	0,7642	1,23	0,8907	1,74	0,9591	2,50	0,9938
0,22	0,5871	0,73	0,7673	1,24	0,8925	1,75	0,9599	2,52	0,9941
0,23	0,5910	0,74	0,7703	1,25	0,8944	1,76	0,9608	2,54	0.9945
0,24	0,5948	0,75	0,7734	1,26	0,8962	1,77	0,9616	2,56	0,9948
0,25	0,5987	0,76	0,7764	1,27	0,8980	1,78	0,9625	2,58	0,9951
0,26	0,6026	0,77	0,7794	1,28	0,8997	1,79	0,9633	2,60	0,9953
0,27	0,6064	0,78	0,7823	1,29	0,9015	1,80	0,9641	2,62	0,9956
0,28	0,6103	0,79	0,7853	1,30	0,9032	1,81	0,9649	2.64	0,9959
0,29	0,6141	0,80	0,7881	1,31	0,9049	1,82	0,9656	2,66	0,9961
0,30	0,6179	0,81	0,7910	1,32	0,9066	1,83	0,9664	2,68	0,9963
0,31	0,6217	0,82	0,7939	1,33	0,9082	1,84	0,9671	2,70	0,9965
0,32	0,6255	0,83	0,7967	1,34	0,9099	1,85	0,9678	2,72	0,9967
0,33	0,6293	0,84	0,7995	1,35	0,9115	1,86	0,9686	2,74	0,9969
0,34	0,6331	0,85	0,8023	1,36	0,9131	1,87	0,9693	2,76	0,9971
0,35	0,6368	0,86	0,8051	1,37	0,9147	1,88	0,9699	2,78	0,9973
0,36	0,6406	0,87	0,8078	1,38	0,9162	1,89	0,9706	2,80	0,9974
0,37	0,6443	0,88	0,8106	1,39	0,9177	1,90	0,9713	2,82	0,9976
0,38	0,6480	0,89	0,8133	1,40	0,9192	1,91	0,9719	2,84	0,9977
0,39	0,6517	0,90	0,8159	1,41	0,9207	1,92	0,9726	2,86	0,9979
0,40	0,6554	0,91	0,8186	1,42	0,9222	1,93	0,9732	2,88	0,9980
0,41	0,6591	0,92	0,8212	1,43	0,9236	1,94	0,9738	2,90	0,9981
0,42	0,6628	0,93	0,8238	1,44	0,9251	1,95	0,9744	2,92	0,9982
0,43	0,6664	0,94	0,8264	1,45	0,9265	1,96	0,9750	2,94	0,9984
0,44	0,6700	0,95	0,8289	1,46	0,9279	1,97	0,9756	2,96	0,9985
0,45	0,6736	0,96	0,8315	1,47	0,9292	1,98	0,9761	2,98	0,9986
0,46	0,6722	0,97	0,8340	1,48	0,9306	1,99	0,9767	3,00	0,9987
0,47	0,6808	0,98	0,8365	1,49	0,9319	2,00	0,9772	3,20	0,9993
0,48	0,6844	0,99	0,8389	1,50	0,9332	2,02	0,9783	3,40	0,9996
0,49	0,6879	1,00	0,8413	1,51	0,9345	2,04	0,9793	3,60	0,9998
0,50	0,6915	1,01	0,8438	1,52	0,9367	2,06	0,9803	3,80	0,9999

Table 2. The first twenty thousand digits of π.
Is the occurrence of the decimals realy random, or do they obey a rule?

$\pi = 3,+$				
1415926535	8979323846	2643383279	5028841971	6939937510
8214808641	3282306647	0938446095	5058223172	5359408128
4428810975	6659334461	2847564823	3786783165	2712019091
7245870066	0631558817	4881520920	9628292540	9171536436
3305727036	5759591953	0921861173	8193261179	3105118548
9833673362	4406566430	8602139494	6395224737	1907021798
0005681271	4526356082	7785771342	7577896091	7363717872
4201995611	2129021960	8640344181	5981362977	4771309960
5024459455	3469083026	4252230825	3344685035	2619311881
5982534904	2875546873	1159562863	8823537875	9375195778
3809525720	1065485863	2788659361	5338182796	8230301952
5574857242	4541506959	5082953311	6861727855	8890750983
8583616035	6370766010	4710181942	9555961989	4676783744
9331367702	8989152104	7521620569	6602405803	8150193511
6782354781	6360093417	2164121992	4586315030	2861829745
3211653449	8720275596	0236480665	4991198818	3479775356
8164706001	6145249192	1732172147	7235014144	1973568548
4547762416	8625189835	6948556209	9219222184	2725502542
8279679766	8145410095	3883786360	9506800642	2512520511
0674427862	2039194945	0471237137	8696095636	4371917287
9465764078	9512694683	9835259570	9825822620	5224894077
4962524517	4939965143	1429809190	6592509372	2169646151
6868386894	2774155991	8559252459	5395943104	9972524680
4390451244	1365497627	8079771569	1435997700	1296160894
0168427394	5226746767	8895252138	5225499546	6672782398
1507606947	9451096596	0940252288	7971089314	5669136867
9009714909	6759852613	6554978189	3129784821	6829989487
5428584447	9526586782	1051141354	7357395231	1342716610
0374200731	0578539062	1983874478	0847848968	3321445713
8191197939	9520614196	6342875444	0643745123	7181921799
5679452080	9514655022	5231603881	9301420937	6213785595
0306803844	7734549202	6054146659	2520149744	2850732518
1005508106	6587969981	6357473638	4052571459	1028970641
2305587631	7635942187	3125147120	5329281918	2618612586
9229109816	9091528017	3506712748	5832228718	3520935396
6711136990	8658516398	3150197016	5151168517	1437657618
8932261854	8963213293	3089857064	2046752590	7091548141
2332609729	9712084433	5732654893	8239119325	9746366730
1809377344	4030707469	2112019130	2033038019	7621101100
2131449576	8572624334	4189303968	6426243410	7732269780
6655730925	4711055785	3763466820	6531098965	2691862056
3348850346	1136576867	5324944166	8039626579	7877185560
7002378776	5913440171	2749470420	5622305389	9456131407

Table 2. cont.

$\pi=3, +$				
6343285878	5698305235	8089330657	5740679545	7163775254
0990796547	3761255176	5675135751	7829666454	7791754011
9389713111	7904297828	5647503203	1986915140	2870808599
8530614228	8137585043	0633217518	2979866223	7172159160
9769265672	1463853067	3609657120	9180763832	7166416274
6171196377	9213375751	1495950156	6049631862	9472654736
6222247715	8915049530	9844489333	0963408780	7693259939
5820974944	5923078164	0628620899	8628034825	3421170679
4811174502	8410270193	8521105559	6446229489	5493038196
4564856692	3460348610	4543266482	1339360726	0249141273
7892590360	0113305305	4882046652	1384146951	9415116094
0744623799	6274956735	1885752724	8912279381	8301194912
6094370277	0539217176	2931767523	8467481846	7669405132
1468440901	2249534301	4654958537	1050792279	6892589235
5187072113	4999999837	2978049951	0597317328	1609631859
7101000313	7838752886	5875332083	8142061717	7669147303
1857780532	1712268066	1300192787	6611195909	2164201989
0353018529	6899577362	2599413891	2497217752	8347913151
8175463746	4939319255	0604009277	0167113900	9848824012
9448255379	7747268471	0404753464	6208046684	2590694912
2533824300	3558764024	7496473263	9141992726	0426992279
5570674983	8505494588	5869269956	9092721079	7509302955
6369807426	5425278625	5181841757	4672890977	7727938000
1613611573	5255213347	5741849468	4385233239	0739414333
5688767179	0494601653	4668049886	2723279178	6085784383
7392984896	0841284886	2694560424	1965285022	2106611863
4677646575	7396241389	0865832645	9958133904	7802759009
2671947826	8482601476	9909026401	3639443745	5305068203
5709858387	4105978859	5977297549	8930161753	9284681382
8459872736	4469584865	3836736222	6260991246	0805124388
4169486855	5848406353	4220722258	2848864815	8456028506
6456596116	3548862305	7745649803	5593634568	1743241125
2287489405	6010150330	8617928680	9208747609	1782493858
2265880485	7564014270	4775551323	7964145152	3746234364
2135969536	2314429524	8493718711	0145765403	5902799344
8687519435	0643021845	3191048481	0053706146	8067491927
9839101591	9561814675	1426912397	4894090718	6494231961
6638937787	0830390697	9207734672	2182562599	6615014215
6660021324	3408819071	0486331734	6496514539	0579626856
4011097120	6280439039	7595156771	5770042033	7869936007
7321579198	4148488291	6447060957	5270695722	0917567116
5725121083	5791513698	8209144421	0067510334	6711031412
3515565088	4909989859	9823873455	2833163550	7647918535

Table 2. cont.

$\pi=3,+$				
6549859461	6371802709	8199430992	4488957571	2828905923
5836041428	1388303203	8249037589	8524374417	0291327656
4492932151	6084244485	9637669838	9522868478	3123552658
2807318915	4411010446	8232527162	0105265227	2111660396
4769312570	5863566201	8558100729	3606598764	8611791045
8455296541	2665408530	6143444318	5867697514	5661406800
1127000407	8547332699	3908145466	4645880797	2708266830
2021149557	6158140025	0126228594	1302164715	5097925923
2996148903	0463994713	2962107340	4375189573	5961458901
0480109412	1472213179	4764777262	2414254854	5403321571
7716692547	4873898665	4949450114	6540628433	6639379003
8888007869	2560290228	4721040317	2118608204	1900042296
4252308177	0367515906	7350235072	8354056704	0386743513
7805419341	4473774418	4263129860	8099888687	4132604721
5695162396	5864573021	6315981931	9516735381	2974167729
4037014163	1496589794	0924323789	6907069779	4223625082
5578297352	3344604281	5126272037	3431465319	7777416031
3162499341	9131814809	2777710386	3877343177	2075456545
3166636528	6193266863	3606273567	6303544776	2803504507
9456127531	8134078330	3362542327	8394497538	2437205835
0408591337	4641442822	7726346594	7047458784	7787201927
8350493163	1284042512	1925651798	0694113528	0131470130
9562586586	5570552690	4965209858	0338507224	2648293972
4803048029	0058760758	2510474709	1643961362	6760449256
2901618766	7952406163	4252257719	5429162991	9306455377
2540790914	5135711136	9410919139	2351910760	2082520261
2784768472	6860849003	3770242429	1651300500	5168323364
1960121228	5993716231	3017114448	4640903890	6449544400
2283345085	0486082503	9302133219	7155184306	3545500766
4611996653	8581538420	5685338621	8672523340	2830871123
2184564622	0134967151	8819097303	8119800497	3407239610
9567302292	1913933918	5680344903	9820595510	0226353536
2711172364	3435439478	2218185286	2608514006	6604433258
0516553790	6866273337	9958511562	5784322988	2737231989
4460477464	9159950549	7374256269	0104903778	1986835938
3634655379	4986419270	5638728317	4872332083	7601123029
0126901475	4668476535	7616477379	4675200490	7571555278
2772190055	6148425551	8792530343	5139844253	2234157623
2276930624	7435363256	9160781547	8181152843	6679570611
3776700961	2071512491	4043027253	8607648236	3414334623
9164219399	4907236234	6468441173	9403265918	4044378051
1266830240	2929525220	1187267675	6220415420	5161841634
9104140792	8862150784	2451670908	7000699282	1206604183

Table 2. cont.

$\pi=3,+$				
9598470356	2226293486	0034158722	9805349896	5022629174
5771028402	7998066365	8254889264	8802545661	0172967026
5178609040	7086671149	6558343434	7693385781	7113864558
6161528813	8437909904	2317473363	9480457593	1493140529
9203767192	2033229094	3346768514	2214477379	3937517034
8244625759	1633303910	7225383742	1821408835	0865739177
3408005355	9849175417	3818839994	4697486762	6551658276
4043523117	6006651012	4120065975	5851276178	5838292041
7267507981	2554709589	0455635792	1221033346	6974992356
7429958180	7247162591	6685451333	1239480494	7079119153
9289647669	7583183271	3142517029	6923488962	7668440323
9588970695	3653494060	3402166544	3755890045	6328822505
6909411303	1509526179	3780029741	2076651479	3942590298
6922210327	4889218654	3648022967	8070576561	5144632064
2248261177	1858963814	0918390367	3672220888	3215137556
2543709069	7939612257	1429894671	5435784687	8861444581
5510500801	9086996033	0276347870	8108175450	1193071412
1596131854	3475464955	6978103829	3097164651	4384070070
3084076118	3013052793	2054274628	6540360367	4532865105
5020141020	6723585020	0724522563	2651341055	9240190274
2645600162	3742880210	9276457931	0657922955	2498872758
4786724229	2465436680	0980676928	2382806899	6400482435
2168895738	3798623001	5937764716	5122893578	6015881617
9906655418	7639792933	4419521541	3418994854	4473456738
3220777092	1201905166	0962804909	2636019759	8828161332
7723554710	5859548702	7908143562	4014517180	6246436267
3114771199	2606381334	6776879695	9703098339	1307710987
7152807317	6790770715	7213444730	6057007334	9243693113
4781643788	5185290928	5452011658	3934196562	1349142415
8584783163	0577775606	8887644624	8246857926	0395352773
2742042083	2085661190	6254543372	1315359584	5068772460
9914037340	4328752628	8896399587	9475729174	6426357455
8798531887	7058429725	9167781314	9699009019	2116971737
3503895170	2989392233	4517220138	1280696501	1784408745
6198690754	8516026327	5052983491	8740786680	8818338510
8282949304	1377655279	3975175461	3953984683	3936383047
2827892125	0771262946	3229563989	8989358211	6745627010
3685406643	1939509790	1906996395	5255300545	0580685501
1920419947	4553859381	0234395544	9597782779	0237421617
8856986705	4315470696	5747458550	3323233421	0730154594
8757141595	7811196358	3300594087	3068121602	8764962867
1465741268	0492564879	8556145372	3478673303	9046883834
9113679386	2708943879	9362016295	1541337142	4892830722

Table 2. cont.

$\pi=3,+$				
1965362132	3926406160	1363581559	0742202020	3187277605
3610642506	3904975008	6562710953	5919465897	5141310348
0861533150	4452127473	9245449454	2368288606	1340841486
5189757664	5216413767	9690314950	1910857598	4423919862
3338945257	4239950829	6591228508	5558215725	0310712570
8475651699	9811614101	0029960783	8690929160	3028840026
7180653556	7252532567	5328612910	4248776182	5829765157
8788202734	2092222453	3985626476	6914905562	8425039127
6407655904	2909945681	5065265305	3718294127	0336931378
7367812301	4587687126	6034891390	9562009939	3610310291
7634757481	1935670911	0137751721	0080315590	2485309066
4366199104	0337511173	5471918550	4644902636	5512816228
1509682887	4782656995	9957449066	1758344137	5223970968
5848358845	3142775687	9002909517	0283529716	3445621296
9748442360	8007193045	7618932349	2292796501	9875187212
3025494780	2490114195	2123828153	0911407907	3860251522
2673430282	4418604142	6363954800	0118002670	4952482017
2609275249	6035799646	9256504936	8183609003	2380929345
4525564056	4482465151	8754711962	1844396582	5337543885
9695946995	5657612186	5619673378	6236256125	2163208628
9279068212	0738837781	4233562823	6089632080	6822246801
0037279839	4004152970	0287830766	7094447456	0134556417
2314593571	9849225284	7160504922	1242470141	2147805734
2339086639	3833952942	5786905076	4310063835	1983438934
7360411237	3599843452	2516105070	2705623526	6012764848
7065874882	2596815793	6789766974	2205700596	8344086973
2162484391	4035998953	5394590944	0704691209	1409387001
4610126483	6999892256	9596881592	0560010165	5256375678
5667227966	1988578279	4848855834	3975187445	4551296563
7634180703	9476994159	7915945300	6975214829	3366555661
4599041608	7532018683	7937023488	8689479151	0716378529
1384987137	5704710178	7957310422	9690666702	1449863746
1963403911	4732023380	7150952220	1068256342	7471646024
0273900749	7297363549	6453328886	9844061196	4961627734
4810688732	0685990754	0792342402	3009259007	0173196036
4495397907	0903023460	4614709616	9688688501	4083470405
2431974404	7718556789	3482308934	1068287027	2280973624
0630792615	9599546262	4629707062	5948455690	3471197299
3831591256	8989295196	4272875739	4691427253	4366941532
0298865925	7866285612	4966552353	3829428785	4253404830
0598135220	5117336585	6407826484	9427644113	7639386692
6977631279	5722672655	5625962825	4276531830	0134070922
7099558561	1349802524	9906698423	3017350358	0440811685

Table 2. cont.

$\pi=3, +$				
2669178352	5870785951	2983441729	5351953788	5534573742
1080485485	2635722825	7682034160	5048466277	5045003126
2432227188	5159740547	0214828971	1177792376	1225788734
2778944236	2167411918	6269439650	6715157795	8675648239
8783419689	6861181558	1587360629	3860381017	1215855272
6435495561	8689641122	8214075330	2655100424	1048967835
3100735477	0549815968	0772009474	6961343609	2861484941
0558642219	6483491512	6390128038	3200109773	8680662877
8936793008	1699805365	2027600727	7496745840	0283624053
6640075094	2608788573	5796037324	5141467867	0368809880
7391253835	5915031003	3303251117	4915696917	4502714943
2342548326	1119128009	2825256190	2052630163	9114772473
1112635835	5387136101	1023267987	7564102468	2403226483
9630878154	3221166912	2464159117	7673225326	4335686146
8850280054	1436131462	3082102594	1737562389	9420751367
4725470316	5661399199	9682628247	2706413362	2217892390
7386480584	7268954624	3882343751	7885201439	5600571048
3604103182	3507365027	7859089757	8272731305	0488939890
9077271307	0686917092	6462548423	2407485503	6608013604
5823626729	3264537382	1049387249	9669933942	4685516483
5220935701	6263846485	2851490362	9320199199	6882851718
5982311225	0628905854	9145097157	5539002439	3153519090
7806101371	5004489917	2100222013	3501310601	6391541589
0594741234	1933984202	1874564925	6443462392	5319531351
9149419394	0608572486	3968836903	2655643642	1664425760
8153740996	1545598798	2598910937	1712621828	3025848112
2994972574	5303328389	6381843944	7707794022	8435988341
1000162076	9363684677	6413017819	6593799715	5746854194
6778113949	8616478747	1407932638	5873862473	2889645643
2961273998	4134427260	8606187245	5452360643	1537101127
5829486191	9667091895	8089833201	2103184303	4012849511
7253558119	5840149180	9692533950	7577840006	7465526031
0425845988	4199076112	8725805911	3935689601	4316682831
7903688786	8789493054	6955703022	6190095020	7643349335
1786227363	7169757741	8302398600	6591481616	4049449650
4434803966	4205579829	3680435220	2770984294	2325330225
5678736400	5366656416	5473217043	9035213295	4352916941
0234529244	9773659495	6305100742	1087142613	4974595615
4595280824	3694457897	7233004876	4765241339	0759204340
3354400515	2126693249	3419672977	0415956837	5355516673
4951827369	5588220757	3551766515	9895519098	6665393549
2254756478	9406475483	4664776041	1463233905	6513433068
4607429586	9913829668	2468185710	3188790652	8703665083

Table 2. cont.

$\pi=3$, +				
8093996270	6074726455	3992539944	2808113736	9433887294
6409089418	0595343932	5123623550	8134949004	3642785271
3610045373	0488198551	7065941217	3524625895	4873016760
8330701653	7228563559	1525347844	5981831341	1290019992
4803118364	4536985891	7544264739	9882284621	8449008777
3343657791	6012809317	9401718598	5999338492	3549564005
5265311709	9570899427	3287092584	8789443646	0050410892
6085902908	1765155780	3905946408	7350612322	6112009373
2008007998	0492548534	6941469775	1649327095	0493463938
7718819682	5462981268	6858170507	4027255026	3329044976
9391760426	0176338704	5499017614	3641204692	1823707648
6683008238	3404656475	8804051380	8016336388	7421637140
2858829024	3670904887	1181900904	9453314421	8287661810
7850171807	7930681085	4690009445	8995279424	3981392135
9239718014	6134324457	2640097374	2570073592	1003154150
4603726341	6554259027	6018348403	0681138185	5105979705
6097164258	4975951380	6930944940	1515422221	9432913021
3151558854	0392216409	7229101129	0355218157	6282328318
3148573910	7775874425	3876117465	7867116941	4776421441
4641766369	8066378576	8134920453	0224081972	7856471983
1865452226	8126887268	4459684424	1610785401	6768142080
2751674573	1891894562	8352570441	3354375857	5342698699
3176085428	9437339356	1889165125	0424404008	9527198378
1194988423	9060613695	7342315590	7967034614	9143447886
0992391350	3373250855	9826558670	8924261242	9473670193
6689511840	0936686095	4632500214	5852930950	0009071510
2611341461	1068026744	6637334375	3407642940	2668297386
3953669134	5222444708	0459239660	2817156551	5656661113
2107119457	3002438801	7661503527	0862602537	8817975194
5780371177	9277522597	8742891917	9155224171	8958536168
0331147639	4911995072	8584306583	6193536932	9699289837
7914710869	9843157337	4964883529	2796328220	7629472823
3890119682	2142945766	7580718653	8065064870	2612289282
0035838542	3897354243	9564755568	4095224844	5541392394
6334893748	4391297423	9143365936	0410035234	3777065888
5987746676	3847946650	4074111825	6583788784	5485814896
4680977870	4464094758	2803487697	5894822824	1239292960
6203534280	1441276172	8583024355	9830032042	0245120728
4461670508	2768277222	3534191102	6341631571	4740612385
7632356732	5417073420	8173322304	6298799280	4908514094
9106024545	0864536289	3545686295	8531315337	1838682656
1173213138	9574706208	8474802365	3710311508	9842799275

Table 2. cont.

$\pi=3, +$				
4426853277	9743113951	4357417221	9759799359	6852522857
4059909935	0500081337	5432454635	9675048442	3528487470
8654489377	6979566517	2796623267	1481033864	3913751865
3471606325	8306498297	9551010954	1836235030	3094530973
8992161255	2559770143	6858943585	8775263796	2559708167
7772370041	7808419423	9487254068	0155603599	8390548985
1345308939	0620467843	8778505423	9390524731	3620129476
7689631148	1090220972	4520759167	2970078505	8071718638
5203802786	0990556900	1341371823	6837099194	9516489600
4639138363	1857456981	4719620184	1080961884	6054560390
3066988317	6833100113	3108690421	9390310801	4378433415
6746964066	6531527035	3254671126	6722246055	1199581831
3936307463	0569010801	1494271410	0939136913	8107258137
0373572652	7922417373	6057511278	8721819084	4900617801
6322439372	6562472776	0378908144	5883785501	9702843779
6029841669	2254896497	1560698119	2186584926	7704039564
1212415363	7451500563	5070127815	9267142413	4210330156
9300806260	1809623815	1613669033	4111138653	8510919367
7086806445	0969865488	0168287434	3786126543	8158342807
9025100158	8827216474	5006820704	1937615845	4712318346
2264482091	0235647752	7230820810	6351889915	2692889108
5904211844	9499077899	9200732947	6805868577	8787209829
6814671769	5976099421	0036183559	1387778176	9845875810
1338413385	3684211978	9389001852	9569196780	4554482858
5726949518	1795897546	9399264219	7915523385	7662316762
6777432242	7680913236	5449485366	7680000010	6526248547
5453076511	6803337322	2651756622	0752695179	1442252808
2754257508	6765511785	9395002793	3895920576	6827896776
8271937831	2654961745	9970567450	7183320650	3455664403
7842645676	3388188075	6561216896	0504161139	0390639601
6464411918	5682770045	7424343402	1672276445	5893301277
0545886056	4550133203	5786454858	4032402987	1709348091
0987399876	6973237665	7370158080	6822904599	2123661689
3216142802	1497633991	8983548487	5625298752	4238730775
2606163364	3296406335	7281070788	7581640438	1485018841
9006664080	6314077757	7257056307	2940049294	0302420498
2827105784	3197535417	9501134727	3625774080	2134768260
8245026792	6594205550	3958792298	1852648007	0683765041
1964965403	2187271602	6485930490	3978748958	9066127250
3223810158	7444505286	6523802253	2843891375	2738458923
3300538621	6347988509	4695472004	7952311201	5043293226
2630971749	5072127248	4794781695	7296142365	8595782090
7557971449	9246540386	8179921389	3469244741	9850973346

Table 2. cont.

$\pi = 3,\ +$				
8409146698	5692571507	4315740793	8053239252	3947755744
2643744980	5596103307	9941453477	8457469999	2128599999
6542616968	5867883726	0958774567	6182507275	9929508931
8172932393	0106766386	8240401113	0402470073	5085782872
1465242385	7762550047	4852964768	1479546700	7050347999
1510102436	5553522306	9061294938	8599015734	6610237122
0277722610	2544149221	5765045081	2067717357	1202718024
4526379628	9612691572	3579866205	7340837576	6873884266
1443545419	5762584735	6421619813	4073468541	1176688311
9467300244	3450054499	5399742372	3287124948	3470604406
3583446283	9476304775	6450150085	0757894954	8931393944
7643800125	4365023714	1278346792	6101995585	2247172201
7235467456	4239058585	0216719031	3952629445	5439131663
9187497519	1011472315	2893267725	3391814660	7300089027
1054967973	1001678708	5069420809	2232908070	3832634534
7550493412	6787613674	6384902063	9640197666	8559232565
3845534372	9141446513	4749407848	8442377217	5154334260
1370924353	0136776310	8491351615	6422698475	0743032971
9637637076	1799191920	3579582007	5956053023	4626775794
8135789400	5599500183	5425118417	2136055727	5221035268
3889710770	8229310027	9766593583	8758909395	6881485602
3624078250	5270487581	6470324581	2908783952	3245323789
8127810217	9912317416	3058105545	9880130048	4562997651
6165356024	7338078430	2865525722	2753049998	8370153487
3938352293	4588832255	0887064507	5394739520	4396807906
5306184548	5903798217	9945996811	5441974253	6344399602
0072629339	5505482395	5713725684	0232268213	0124767945
4555711266	0396503439	7896278250	0161101532	3516051965
0135295661	3978884860	5097860859	5701773129	8155314951
4466283998	8060061622	9848616935	3373865787	7359833616
4837011709	6721253533	8758621582	3101331038	7766827211
7547570354	6994148929	0413018638	6119439196	2838870543
3055861598	9991401707	6983854831	8875014293	8908995068
1651716677	6672793035	4851542040	2381746089	2328291703
4453184040	4185540104	3513483895	3120132637	8369283580
4490453627	5600112501	8433560736	1222765949	2783937064
6202215368	4941092605	3876887148	3798955999	9112099164
8158686952	5069499364	6101756850	6016714535	4315814801
0556211671	5468484778	0394475697	9804263180	9917564228
0259627304	3067931653	1149401764	7376938735	1409336183
5955595546	5196394401	8218409984	1248982623	6737714672
1431885988	2769449011	9321296827	1588841338	6943468285
4165654797	3670548558	0445865720	2276378404	6682337895

Table 2. cont.

$\pi=3,+$				
4502285157	9795797647	4670228409	9956160156	9108903845
8365620945	5543461351	3415257006	5974881916	3413595567
7948282769	3895352178	3621850796	2977851461	8843271922
8442253547	2653098171	5784478342	1582232702	0690287232
6282727632	1779088400	8786148022	1475376578	1058197022
8307332335	6034846531	8730293026	6596450137	1837542889
2679332107	2686870768	0626399193	6196504409	9542167627
1591845821	5625181921	5523370960	7483329234	9210345146
3996122816	1521931488	8769388022	2810830019	8601654941
8052187292	4610867639	9589161458	5505839727	4209809097
4627134946	3685318154	6969046696	8693925472	5194139929
5888676950	1612497228	2040303995	4632788306	9597624936
3547891129	2547696176	0050479749	2806072126	8039226911
2968106203	7765788371	6690910941	8074487814	0490755178

Name Index

Subject Index